ALS

Advances in Life Sciences

Water and Ions
in Biomolecular Systems

Proceedings of
the 5th UNESCO International Conference

Edited by
D. Vasilescu
J. Jaz
L. Packer
B. Pullman

1990

Birkhäuser Verlag
Basel · Boston · Berlin

Editors' addresses:

Prof. D. Vasilescu
Laboratoire de Biophysique
Université de Nice
Parc Valrose
06034 Nice Cedex
France

Prof. L. Packer
Membrane Bioenergetic Group
Lawrence Berkeley Laboratory
University of California
Berkeley, CA 94720
USA

Prof. J. Jaz
UNESCO
Place de Fontenoy
75700 Paris
France

Prof. B. Pullman
Institut de Biologie
Physico-Chimique
13, rue Pierre et Marie Curie
75005 Paris
France

Cover illustration by Biophysics Laboratory, Nice.

Deutsche Bibliothek Cataloguing-in-Publication Data
Water and ions in biomolecular systems: proceedings of the ... UNESCO international conference. – Basel; Boston; Berlin: Birkhäuser.
(Advances in life sciences)
Bis 4 (1988) u.d.T.: Water and ions in biological systems
NE: Unesco
5 (1990)
ISBN 3-7643-2359-0 (Basel...)
ISBN 0-8176-2359-0 (Boston)

Printed in Germany on acid-free paper
ISBN 3-7643-2359-0
ISBN 0-8176-2359-0

CONTENTS

MISCELLANEOUS

5

FOREWORD

About 80 scientists from all over the world have gathered at the Parc Valrose Scientific Campus of Nice - Sophia Antipolis University, for the 5th UNESCO International Conference on "Water and Ions in Biomolecular Systems", to present and discuss current problems in the field.

The Conference proceedings comprise selected and reviewed contributions of invited plenary lectures, lectures and posters presentations which cover all the Biophysical sectors of research on the following three major topics :

* Water, Ions and Nucleic Acids

* Water, Ions and Membranes

* Water, Ions and Proteins

The organizers of the 5th UNESCO International Conference on "Water and Ions in Biomolecular Systems" have obtained financial support from various sponsors. They would like to express their appreciation and gratitude to these organizations for the valuable moral and financial support which has made the arrangement of these Conference in Nice possible.

The Meeting has been sponsored by :

* United Nations Educational Scientific and Cultural Organization
* The University of Nice - Sophia Antipolis
* D.R.E.T. (Defence Ministery)
* Conseil Général des Alpes Maritimes
* Ville de Nice
* The International Union of Biochemistry
* The International Union of Pure and Applied Biophysics
* Banque Nationale de Paris
* Air Liquide
* Nice Congrès

The Executive Committee, Professors J. JAZ, B. PULLMAN and D. VASILESCU, would like also to thank the Members of our Scientific Committee for valuable comments and suggestions which have greatly helped the organization of the 1989 UNESCO International Conference. Administrative and technical help provided by the staff of the Biophysics Laboratory of the University of Nice - Sophia Antipolis is gratefully aknowledged. Concerning the editorial work, our special thanks go to Mr. A. BALLY (Biology Editor) and Mrs C. JOYCE (Editorial Assistant) from BIRKHAUSER Verlag and to Mrs N. GASIGLIA (Biophysics Laboratory, Nice).

Dan VASILESCU
Conference Coordinator

PREFACE

Since the very creation in 1976, at a meeting in Budapest, of UNESCO European Expert Committee on Molecular Biophysics, the theme of "Water and Ions in Biomolecular Systems" has been one of the dominating topics of creative collaboration between the nations and scientists engaged in or associated with the activity of this very efficient Committee. The theme was moreover unique in the sense that it was the only one which has been the subject of regular international conferences carried out under the sponsorship of the Committee. The four first Conferences were held in Bucarest, Rumania. For a number of reasons it appeared appropriate to hold the fifth one in western Europe and Nice was chosen for this sake, to a large extend due to the availability at its University of a volontary and devoted organizer, Professor Dan Vasilescu.

One of the obvious advantages of a series of related meetings, held at regular intervals on the same theme, is to enable a continuous and critical evaluation of the developments which occur in this theme. When the subject is of the importance and universality of the one considered here this evaluation becomes particularly important.

This volume contains the Proceedings of the 5th Conference. Its comparison with the Proceedings of the previous, in particular the earliest ones, demonstrates the remarkable progress during the last decade in our knowledge of the role of water and ions in biological systems, due to a large extent to the striking progress of both experimental and theoretical techniques for studying this problem. Considered initially as essentially environmental factors, water and ions have now been recognized to form, to a measurable extent, a constituant part of the essential biomolecular systems. Also, the largely static viewpoint which dominated at the first conferences has now yielded ground to a more dynamic view of these associations.

As testified by the content of this volume, the meeting was exceedingly rich in remarkable contributions which covered extensively the three main divisions in which the role of water and ions was considered, namely : nucleic acids, proteins and membranes. A good equilibrium was obtained thanks to a careful selection of the main speakers. The volume which provides an excellent summing up of the present knowledge on the subject should thus be of utmost interest and usefullness to biochemists and biophysicists who at one or another stage of their researches are directly interested in the explicit role of water and ions. And is there anybody who is not ?

The UNESCO European Expert Committee on Molecular Biophysics is in the process of being absorbed, under the instructions of the new Director General of UNESCO, Professor Federico Mayor, into a more ambitious, world-wide UNESCO Committee on Molecular and Cell Biology. There is, however, no doubt that the theme of "Water and Ions in Biomolecular Systems" will remain as an important object of studies of this new organization.

The Conference took place in the beautiful site of the University of Nice - Sophia Antipolis. The University and the City provided a most gracious hospitality. But, above all, the merit of the success of the Conference and of the speedy presentation of this volume must be attributed to its main organizer, Professor Dan Vasilescu who deserves for this achievement the praise and the gratitude of the scientific Community. Thanks are also due to Professor José Jaz, former Director of the European Office of Scientific Cooperation of UNESCO, for his devoted role as the representative of the Director General of UNESCO and efficient coorganizer of the Conference.

Bernard PULLMAN
Honorary Chairman of the UNESCO European Expert
Committee on Molecular Biophysics

WATER, IONS AND NUCLEIC ACIDS

STRUCTURAL WATER OF NUCLEIC ACIDS

E. WESTHOF

Institut de Biologie Moléculaire et Cellulaire, Centre National de la Recherche Scientifique, 15 rue R. Descartes, F-67084 Strasbourg-Cedex.

SUMMARY : Nucleic acids, through variations in torsion angles of the sugar-phosphate backbone and through reorientations of the bases, can adapt their structures so that their polar hydrophilic atoms form three-dimensional networks able to interact favorably with the molecules of the solvent. This interdependence between solvent and nucleic acid structure forms the physicochemical basis for DNA polymorphism. Around unusual sugar-phosphate backbone conformations and non-canonical base pairs there is an increased frequency of 3'-phosphate-water-base and 3'-phosphate-water-sugar bridges as well as base-water-base intra- or intermolecular water bridges. This observation stresses the stabilizing roles and structural importance of water bridges in non-standard conformations and/or base pairings. It is suggested that the dynamics of the hydration structure around helical nucleic acids in aqueous solution is only slightly perturbed, except for the structurally important water fraction around repetitive sequences (spines of A-T stretches), unusual conformations (phosphate-water-base bridges of syn bases), or non-Watson-Crick base pairs (G-T, A-A, G-A,...).

INTRODUCTION

In nucleic acids, tertiary structure is considered to be the result of an equilibrium between electrostatic forces due to the negatively charged phosphates, stacking interactions between the bases due partially to hydrophobic and dispersion forces, hydrogen bonding interactions between the polar substituents of the bases, and the conformational energy of the sugar-phosphate backbone. In its preferred conformations, the polynucleotide backbone exposes the negatively charged phosphates to the dielectric screening by the solvent and promotes the stacked helical arrangement of adjacent bases. In this way, a hydrophobic core is created where hydrogen bond formation between bases as well as additional sugar-base and sugar-sugar interactions are favored. In such helical structures, only the internal atoms involved in hydrogen bonding between the bases are protected from solvent with most of the other atoms accessible to water. Thus, water molecules contribute to the stability of helical conformations of nucleic acids by screening the charges of the phosphates, by bonding to the polar exocyclic atoms of the bases, and by influencing the conformations of residues with methyl groups via hydrophobic interactions. Due to the periodicity of the helical structures of nucleic acids, water sites and bridges involving the polar base atoms lead to structured arrangements of water molecules, called columns, chains, filaments (Clementi & Corongiu, 1981), or spines (Drew & Dickerson, 1981).

The plausibility that hydration plays a role in the stability of nucleic acid helices was first suggested by Geiduschek & Gray (1956). Later, base stacking forces together with hydrogen bonding between complementary bases were held responsible for double helical structures in solution. In 1967, Lewin propounded the concept that water-bridges contribute greatly to the stability of the DNA double helix in solution on the basis of model building and theoretical considerations. However, the evidence was indirect and the lack of direct crystallographic experimental evidence held the paper in respectable obscurity.

Recently, extended reviews have appeared on nucleic acid hydration stressing the importance of water molecules and water-bridges in nucleic acid stability on the basis of crystallographic data (Saenger, 1987; Westhof, 1987, 1988) and simulation results (Westhof & Beveridge, 1990). Here, it will be emphasized that similar water binding sites and water bridges are found repeatedly in small as well as large nucleic acids and that they play an important part in the stability of non-standard conformations and/or base pairings.

<u>HYDRATION AROUND PHOSPHATE GROUPS</u> : Around each anionic phosphate oxygen, the motif most frequently seen, in nucleotides as well as in oligonucleotides, is the "cone of hydration" with its three water molecules. This motif was predicted by Pullman and coworkers using quantum mechanical calculations (Langlet, et al. 1979). Recently, Monte Carlo simulations of the B-dodecamer d(CGCGAATTCGCG) revealed the extensive presence of "cones of hydration" around anionic phosphate oxygen atoms (Subramanian & Beveridge, 1989). Crystallographically, "cones of hydration" are also seen in the mentionned B-dodecamer (Westhof & Beveridge, 1990). Another example coming from the Z-DNA hexamer d(5BrCG5BrCG5BrCG) (Chevrier, et al. 1986) is shown below. The crystal structure of inosine-5'-monosphosphate (Rao & Sundaralingam, 1969) displays also two "cones of hydration" around the phosphate group. Up to now, such "cones of hydration" have not be seen in structures of RNA molecules larger than the nucleotides. This is most probably due to lack of resolution. However, in several dinucleotide structures, the binding of sodium ions perturbs such arrangements, since direct ion binding and through-water binding are of the same order of magnitude (Pullman, et al. 1978). Instead, helical RNA structures (or A-DNA type structures of deoxyoligomers) present frequently a water bridge between anionic phosphate oxygen atoms of successive phosphate groups on the same strand (Saenger, et al. 1986; Westhof, et al. 1988).

<u>3'-PHOSPHATE-WATER-SUGAR BRIDGES</u> : In RNA structures, a water bridge between the O2' hydroxyl group and an anionic oxygen atom of the 3'-phosphate group is seen only when the sugar-phosphate backbone makes sharp turns and adopts non-helical conformations. Two examples are shown below. In one drawing, the phosphate between A58 and U59 of yeast tRNA-asp adopts the <u>trans-gauche plus</u> conformation; and in the other, the phosphate between C20 and A21 adopts the <u>gauche minus-gauche plus</u> conformation. The structural and stabilizing importance of such water bridges should be kept in mind when computing and model building unusual conformations of nucleic acids.

<u>5'-PHOSPHATE-WATER-BASE BRIDGES</u> : In DNA structures, such bridges are not frequent, except between methyl groups of thymines and their attached phosphates (Drew & Dickerson, 1981). Water bridges (with one or two water molecules) between the N7 atoms of purines and their attached 5'-phosphate are very frequent in A-form DNA structures or RNA structures (Westhof, 1987).

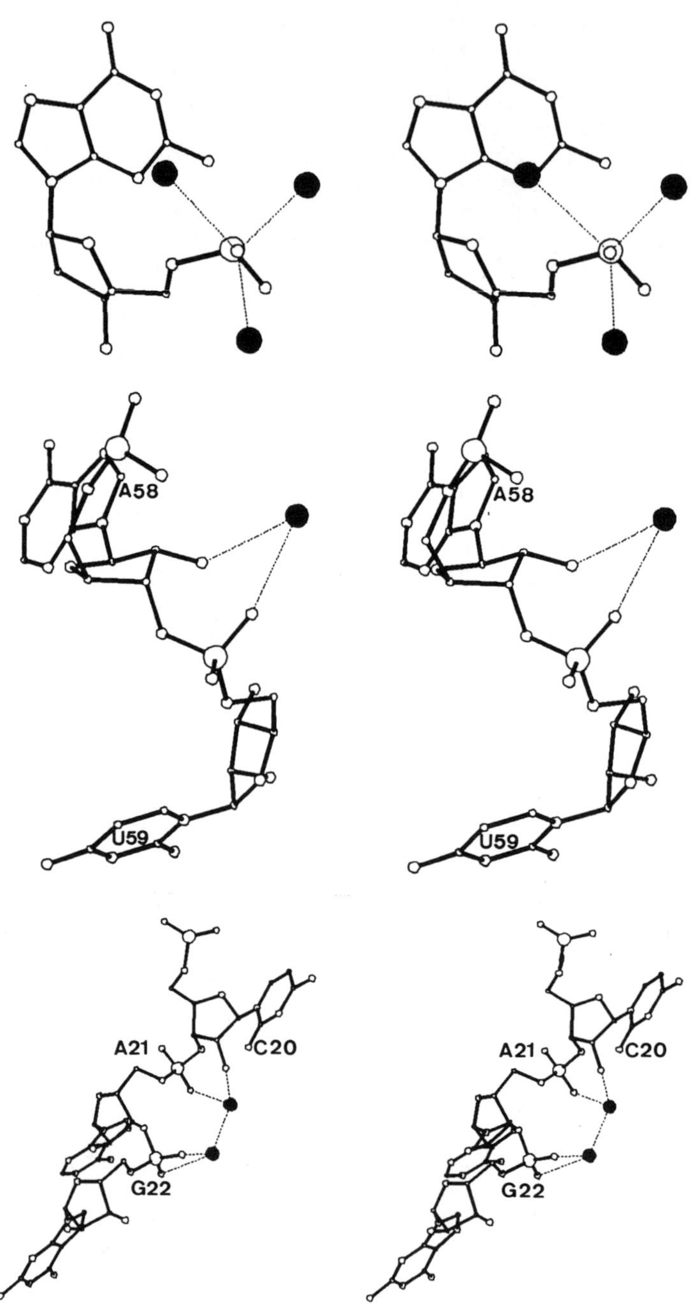
A58
U59
A21
C20
G22

3'-PHOSPHATE-WATER-BASE BRIDGES : In right-handed helical structures (either RNA or DNA), such bridges are not observed. However, around non-helical conformations or non-canonical base pairs, the frequency and structural role of 3'-phosphate-water-bridges increase. Thus, the syn conformation of the guanine base in Z-DNA is stabilized by an intramolecular water bridge between its amino group N2 and one of its anionic phosphate oxygens (Wang, et al. 1979). Such a situation was seen also in the crystal structure of the nucleotide 2-amino-8-methyl-adenosine 5'-monophosphate (Silverton, et al. 1982) in which the base adopts the syn conformation about the glycosyl bond (see left drawing below). Similarly, in the Ganti-Asyn base pair inserted into a B-dodecamer, there is a water molecule linking the adenine N3 and one of its 3'-anionic phosphate oxygen (Hunter, et al. 1986).

Besides additional base-water-base bridges across the pair (Ho, et al. 1985; Kennard, 1986; Westhof, et al. 1988), unusual base pairs present also new 3'-phosphate-water-base bridges. Thus, in A-A self-pairs, the amino N6 of one adenine is linked via a water molecule to an anionic phosphate oxygen of the opposite base (see right drawing below; from Westhof, et al. 1980) In the ApApA structure, the N6 atom is linked directly to the opposite phosphate with the N1 atom linked to the same phosphate group via a water molecule (Suck, et al. 1976). A similar situation occurs in the adenosine-proflavine sulfate complex where sulfate groups play the role of phosphates (Swaminathan, et al. 1982).

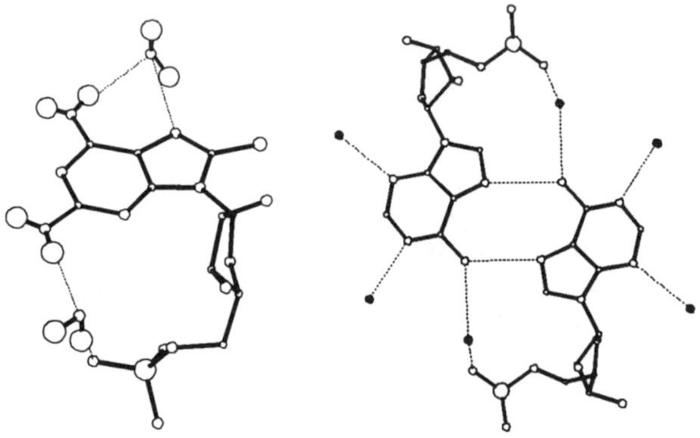

Two examples of similar hydration patterns around the
A9-A23 pair and the G22-A46 pair in yeast tRNA-asp are
also shown below (Westhof, et al. 1988).

SUGAR-WATER-BASE BRIDGES : Such bridges occur in the
minor groove of helical nucleic acids. They are the least
dependent on the nucleic acid form (A, B, or Z), on the
sequence, and on the nature of the sugar. Water molecules
participating in such bridges are tightly bound and appear
clearly in electron density maps. They should thus
contribute to the stability of all helical nucleic acids.
The most frequent sugar-water-base bridge in RNA
structures links the O2' hydroxyl group to the exocylic O2
atom of pyrimidines or to the ring nitrogen N3 of purines
(Westhof, 1988). In DNA structures, the O2/N3...W...O4'
bridge within the same residue occurs but is rare (example
in the decamer structure d(CCAAGATTGG), Privé, et al.
1986). A theoretical study found such bridging water
molecules (Perahia, et al. 1977). Otherwise, in DNA and
in RNA, the O2/N3...W...O4' water bridge occurs between
successive residues on the same strand. The sequences
rich in A-T's, together with the helical periodicity, lead
to spines of hydration running down the central canyon of
the minor groove (Drew & Dickerson, 1981).

BASE-WATER-BASE BRIDGES : Such bridges, when
intra-residue, occur only around purine bases; in the
major groove, between N6/O6 and N7 and, in the minor
groove, between N2 and N3. Inter-residue base-water-base
bridges, inter-strand as well as intra-strand, are
versatile and variable especially in the major groove (see
Westhof, 1987). Their localizations in electron density
maps is often patchy, revealing not only their mobility
but also the protean nature of their arrangements around
the hydrophilic sites of the major groove of helical
nucleic acids. Those around unusual base pairs are often
very clearly seen in electron density maps.

CONCLUSIONS

As already advocated by Lewin (1967), water-bridges at different sites of the nucleic acids appear not to be of the same strength. The main intra-nucleotide water bridges are schematically represented below with the thickness of the lines representing roughly the frequency of occurrence. Strong ones appear mainly in the minor groove of helical nucleic acids and should contribute significantly to the stability of the helical arrangements of nucleic acid in aqueous solution. The 3'-phosphate-water-base and 3'-phosphate-water-sugar water bridges around unusual nucleotide conformations (e.g. syn conformation of the base or the gauche minus-gauche plus conformation of the sugar-phosphate backbone) and around non-canonical base pairs (e.g. A-G, A-A U-G pairs) are crucial for the stability of such conformations. The critical role of the sugar-water-base bridges in the minor groove contrasts with the versatility and mobility of the base-water-water bridges and arrangements in the major groove of helical nucleic acids. In short, water molecules constitute an integral part of helical as well as non-helical structures of nucleic acids.

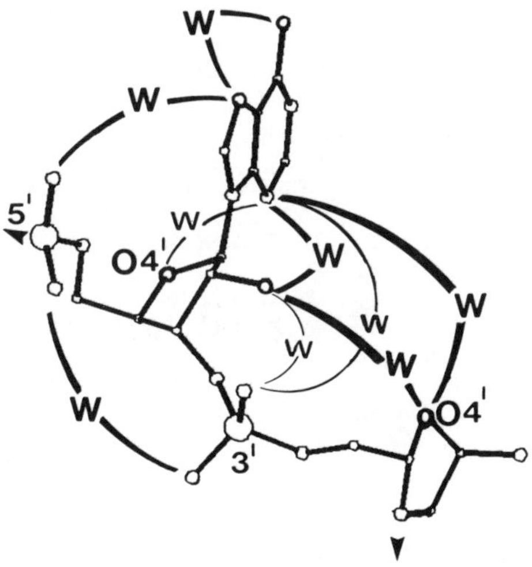

REFERENCES

Chevrier, B., Dock, A.C., Hartmann, B., Leng, M., Moras, D., Thuong, M.T., & Westhof, E. (1986) J. Mol. Biol. 188, 707-719.

Clementi, E & Corongiu, G. (1981) In: Biomolecular Stereodynamics, vol.1 (R.H. Sarma, Ed.), Adenine Press, New York, pp. 209-259.

Drew, H.R. & Dickerson, R.E. (1981) J. Mol. Biol. 151, 535-556.

Geiduschek, E.P. & Gray, I. (1956) J. Amer. Chem. Soc. 78, 879-880.

Ho, P.S., Frederick, C.A., Quigley, G.J., van der Marel, G.A., van Boom, J.H., Wang, A.H.-J., & Rich, A. (1985) EMBO J. 4, 3617-3623.

Hunter, W.N., Brown, T., & Kennard, O. (1986) J. Biomol.Struct. Dyn. 4, 173-191.

Kennard, O. (1986) J. Biomol.Struct. Dyn. 3, 205-226.

Langlet, J., Claverie, P., Pullman, B., & Piazzola, D. (1979) Int. J. Quant. Chem. QBS 6, 409-437.

Lewin, S. (1967) J. Theoret. Biol. 17, 181-212.

Privé, G.G., Heinemann, U., Chandrasegaran, S., Kan, L.S., Kopka, M.L., & Dickerson, R.E. (1987) Science 238, 498-504.

Pullman, A., Pullman, B., & Berthod, H. (1978) Theoret. Chim. Acta (Berl.) 47, 175-192.

Rao, S.T. & Sundaralingam, M. (1969) J. Amer. Chem. Soc. 91, 1210-1217.

Saenger, W. (1987) Ann. Rev. Biophys. Biophys. Chem. 16, 93-114.

Saenger, W. , Hunter, W.N., & Kennard, O. (1986) Nature 324, 385-388.

Silverton, J.V., Limm, W., & Miles, H.T. (1982) J. Amer. Chem. Soc. 104, 1082-1087.

Subramanian, P.S. & Beveridge, D.L. (1989) J. Biomol. Struct. Dyn. 6, 1093-1122.

Suck, D., Manor, P.C., & Saenger, W. (1976) Acta Cryst. B32, 1727-1737.

Swaminathan, P., Westhof, E., & Sundaralingam, M. (1982) Acta Cryst. B38, 515-522.

Wang, A.H.-J., Quigley, G.J., Kolpak, F.J., Crawford, J.L., van Boom, J.H., van der Marel, G.A., & Rich, A. (1979) Nature 282, 680-686.

Westhof, E. (1987a) Int. J. Biol. Macro. 9, 186-192.

Westhof, E. (1987b) J. Biomol. Struct. Dyn. 5, 581-600.

Westhof, E. (1988) Ann. Rev. Biophys. Biophys. Chem. 17, 125-144.

Westhof, E. & Beveridge, D.L. (1990) Water Science Reviews 5, in press.

Westhof, E., Dumas, P., & Moras, D. (1988) Biochimie 70, 145-165.

Westhof, E., Rao, S.T., & Sundaralingam, M. (1980) J. Mol. Biol. 142, 331-361.

Water and Ions in Biomolecular Systems
Advances in Life Sciences
© 1990 Birkhäuser Verlag Basel

STRUCTURE OF <u>XENOPUS LAEVIS</u> 5S rRNA AS DETERMINED BY SOLUTION DATA
AND COMPUTER GRAPHIC MODELING.

P. Romby, C. Brunel, E. Westhof, C. Ehresmann and B. Ehresmann.

Institut de Biologie Moléculaire et Cellulaire du CNRS, 15 rue R.
Descartes, 67084 Strasbourg-Cedex, France.

SUMMARY: A detailed atomic model of the eukaryotic oocyte 5S rRNA
from <u>Xenopus laevis</u> has been built using computer modeling. The
model integrates stereochemical constraints and experimental data
on the accessibility of bases and phosphates towards several
structure-specific probes. In spite of some uncertainty in the
relative orientation of the substructures, several conclusions
have been reached: (i) the model adopts a distorted Y-shaped
structure; (ii) no long-range tertiary interaction occurs but the
internal loops adopt particular conformation containing non-
canonical base pairs. In this paper, the model is corroborated by
the structural analysis of several mutants of loop c.

Ribonucleic acids molecules are involved in a wide range of
functions in the cell. Obviously, their biological activities are
essentially governed by their three-dimensional structure and
their ability to assume conformational switches. X-ray diffraction
is a powerfull method for the tertiary structure determination.
However, only a few RNA crystal structures are known at high
resolution. In the last years, experimental approaches essentially
based on the use of chemical and enzymatic probes were devised in
order to study the conformation of RNAs in solution under nearly
physiological conditions (Ehresmann et al., 1987). Furthermore,
this approach provides specific information on interactions of the
RNA with other macromolecules and may reveal conformational
rearrangements.

Recently, we have probed at nucleotide resolution the
secondary and tertiary structure of spinach chloroplast 5S rRNA
(Romby et al., 1988) and of the oocyte and somatic 5S rRNAs from
<u>Xenopus laevis</u> (Romaniuk et al., 1988) by the use of a large
variety of structure-specific probes. Three-dimensional structure

models of these 5S rRNAs were built by graphic modeling (Westhof et al., 1989). These models integrate experimental data and stereochemical constraints. In order to test and to refine the conformation of the different regions of the _X. laevis_ 5S rRNA, several residues have been selected for site-directed mutagenesis. The effect of the mutation on the conformation of the 5S rRNA has been tested. In this paper, we will focused our discussion on the conformation of loop c.

MATERIAL AND METHODS

Preparation of 5S rRNA: The wild type oocyte 5S rRNA and the mutants were obtained from cloned genes by _in vitro_ transcription with T7 RNA polymerase (Romaniuk et al., 1987).

Limited enzymatic digestions and chemical modifications: The different enzymes and chemicals used in this work are listed in Table 1. The RNA is subjected to the limited action of the probe,

probes	molecular weight	specificity
Enzymes		
RNase T1	11,000	unpaired G
nuclease S1	32,000	unpaired N
RNase V1	15,900	paired or stacked N
Chemicals		
DMS	126	N1-A>N3-C,N7-G
DEPC	174	N7-A
CMCT	424	N3-U>N1-G
ENU	117	phosphate oxygens

Table 1: Chemical and enzymatic probes (from Ehresmann et al., 1987). The molecular weight and the specificity of the various probes are given.

in such a way that less than one cut or modification is statistically introduced per molecule. Particular attention must be kept to the fact that probing experiments have to be conducted under strictly defined conditions, since RNA conformation is sensible to ionic environment, pH and temperature. Chemical modifications are conducted under three different conditions: native conditions (corresponding to optimum functional conditions, at 20°C and in the presence of 5 mM MgCl2 and 100 mM KCl), semi-denaturing conditions (at 20°C, in the presence of 1 mM EDTA) and denaturing conditions (at 90°C, in the presence of 1 mM EDTA).

Graphic modeling: The building of a tertiary model is based first on the recognition of the elementary motifs constituting the secondary structure, followed by the choice of the most appropriate motifs in a structural bank and then by the assembly of these elements into a tertiary model following stereochemical rules. In the structure bank are held the tertiary structure of crystallographic determined nucleic acid structures as well as of already modeled structures. Model building is assisted by several computer programs (Westhof et al., 1989). The various steps and programs for the modeling are listed in Table II.

I. Construction of the secondary structure and data bank containing the phylogenetic and experimental results by PROT (Amerein, 1988).

II. Construction of the tertiary fragments according to the sequence and the secondary structure either from standard helices by NAHELIX or from the structure data bank by FRAGMENT (Westhof, 1988).

III. Manipulation of each fragment for unusual base pairing or bulged residue and assembly of the fragments by SAM/FRODO (Jones, 1978).

IV. Refinement of the fragments and of the assembled structure with geometrical and stereochemical restraints by NUCLIN/NUCLSQ (Westhof et al., 1985).

V. Energy minimization by AMBER (Weiner and Kollman, 1981).

VI. Computation of the accessibilities by ACCESS (adapted from Richmond, 1984).

VII. Go to III until agreement with experimental data.

Table II: Computer modeling of nucleic acids (from Westhof,1988).

At the end of the modeling as well as during the construction of the sub-structures, the accessible surface of the probed atoms are calculated and checked against the experimental results. The radius of a sphere rolling onto the van der Waals sphere is adapted to the chemical probe used (Holbrook and Kim, 1983; Lavery and Pullman, 1984).

RESULTS AND DISCUSSION

The tertiary model of Xenopus laevis 5S rRNA: The experimental data (Romaniuk et al., 1987) confirm the existence of the five helices predicted in the consensus model derived from sequence comparison, and do not support the existence of tertiary interactions between the three arms of the RNA (Fig. 1). The deduced tertiary model adopts a Y-shape with a short stalk made of stem 1 and with the two arms of the Y made of stems 2 and 3 (Fig. 2). Helices B and D are stacked and not far from colinearity. The orientation of the stalk with respect to the arms appears to be controlled by residues of loop a (Westhof et al., 1989).

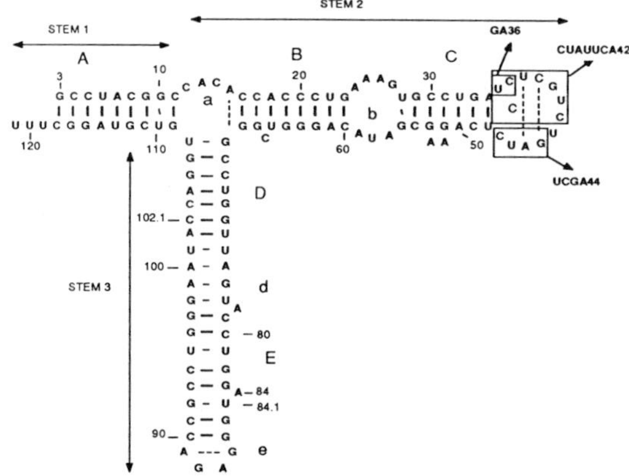

Figure 1: Secondary structure of *Xenopus laevis* oocyte 5S rRNA showing the different mutation sites in loop c. The numbering is according to Westhof et al. (1989).

The regions depicted as unpaired internal loops appear to fold into complex and organized structures. In particular, an unusual structure for region d (nucleotides 74.1-78/98-101) is found in which non-canonical A-A, A-G and U-U base pairs are postulated and several phosphates involved in cation binding site have been identified.

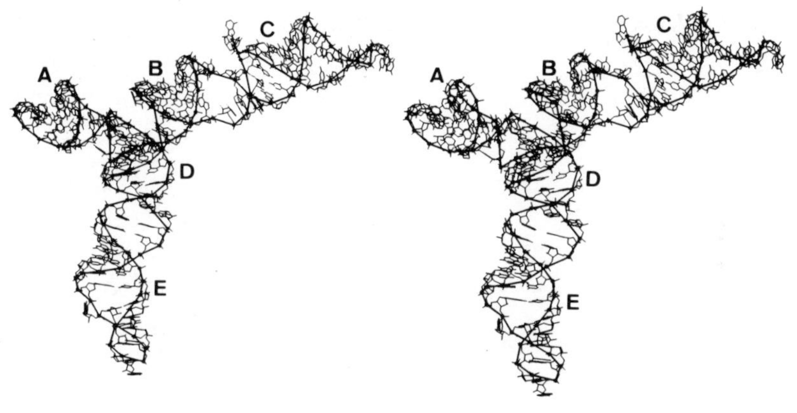

Figure 2: Three-dimensional model of _Xenopus laevis_ oocyte 5S rRNA. The phosphate backbone is shown by heavy lines. All stereoviews were drawn with the program PLUTO (S. Motherwell and P. Evans, MRC Cambridge).

Loop c in X. laevis 5S rRNA: An intrinsic conformation of this loop is suggested by probing data (Fig. 3a,b). The absence of reactivity of U37(N3), C38(N3), G44(N1) and A45(N7) suggests the existence of a Watson-Crick base pair between C38 and G41 and of a reverse Hoogsteen pair between U37(N3,O2) and _trans_ A45(N7,N6). The two pyrimidines on each side of loop c are inside the loop with C36 and C47 stacked to explain the protection of their N3 atoms, but without H-bonds. The loop is closed by turning between G39 and U40 with U40 to U43 being in an anticodon-like conformation (Fig. 3c). Nucleotides GUCU43 point in the direction of the major groove of helix C.

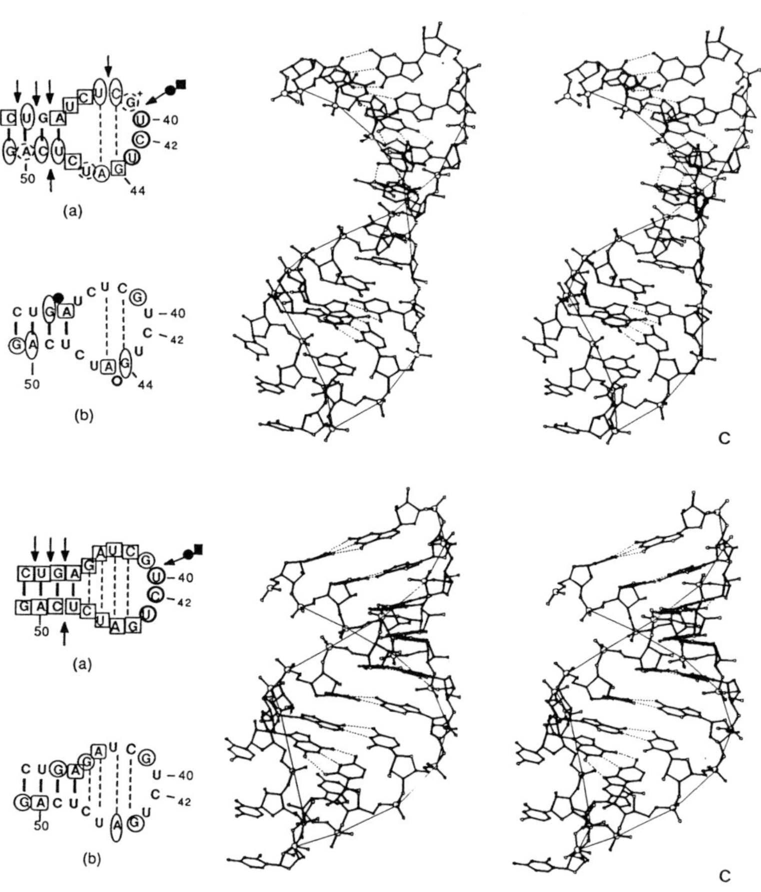

Figure 3: (Top) Tertiary folding of loop c of X. laevis oocyte 5S
rRNA. (a) Accessibility towards enzymatic probes: (→) RNase V1,
(●→) nuclease S1, (■→) RNase T1. Reactivity of Watson-Crick
positions: reactive under native conditions, (●) strong hit,
(O) moderate hit and (○) marginal hit; unreactive under native

conditions but reactive under semi-denaturing conditions, (O);
(O +) denotes increased reactivity under semi-denaturing
conditions, as compared to native conditions; unreactive under
both native and semi-denaturing conditions (□). (b) Reactivity
of N7 position of purines: same symbols as in a. Reactivity of
phosphates: (●,O) strong and moderate protection, respectively.
(c) Stereoscopic views of loop c (residues C31 to G51). The
phosphate backbone is shown by thin lines and the H-bonds by
broken lines. (Bottom) Tertiary folding of loop c of _Xenopus
laevis_ oocyte 5S rRNA GA36 mutant. Same legend as above (from
Brunel et al., to be published).

The replacement of UC36 by GA36 led to the prolongation of
helix C and to the formation of a four-membered loop (Fig. 4a).
The fact that nucleotides 31-38/44-51 in mutant GA36 become
reactive only under denaturing conditions suggests a higher
stability of this region in the mutant as compared to the wild
type (Fig. 4a,b). In agreement with this observation, it has been
shown that hairpin loops of four or five nucleotides are the most
stable (Groebe and Uhlenbeck, 1988). Interestingly, the
reactivity pattern of nucleotides 39 to 43 is identical to that
of the wild type (Fig. 3 and 4).

In another mutant, residues GAUC47 were substituted by
AGCU47. Probing experiments led us to propose two Watson-Crick
base pairs between G45-C38 and A44-U39 with the two cytosines 38
and 46 in a bulged out conformation. In this case also, the
reactivity pattern of residues G39 to U43 remains identical to
that of the wild type. This suggests that the four-membered loop
motif GUCU43 with the G39-turn is also found in this mutant.
However, the four-membered loops in the two mutants and in the
wild type point in different orientations.

When UCUCGUC42 is replaced by CUAUUCA42, all nucleotides
become exposed to the solvent and the loop is easily cleaved by
nuclease S1 suggesting the absence of any particular
conformation. In mutants AGCU47 and CUAUUCA42, the stability of
helix C is not affected. Remarkably, the different mutations
conducted in loop c result only in a local rearrangement but do
not perturb the rest of the molecule.

CONCLUSION

The present paper shows that biochemical approaches, coupled with computer graphic modeling and mutagenesis represent a powerful approach to investigate the conformation of RNA at atomic level. Our data confirm that the different domains of the 5S rRNA are independent and that no tertiary interaction occur between loop c and region d or loop e as it was previously suggested (Pieler and Erdmann, 1982; Toots et al., 1982). Our results also stress the importance of nucleotides U37, C38, G44 and A45 in maintaining the intrinsic conformation of loop c. Also, it appears that long hairpin loops are less stable than hairpin loops of four residues. Hence, such long hairpin loop could display dynamic properties that should greatly facilitate interaction with other macromolecules. Indeed, several of the residues of loop c have been implicated in protein binding (Christiansen et al., 1987).

Experiments on several mutants in regions b, d and e are now in progress in order to refine the conformation of these regions and to define their role in the tertiary structure. The co-axiality between helices B and D will be also tested by mutants in loop a.

ACKNOWLEDGEMENTS

We are grateful to Prof. P. Romaniuk for the preparation of the RNA transcripts and for helpful discussions. We thank F. Baudin for stimulating discussions and Prof. J.P. Ebel for constant interest and support. The graphic modeling was done on an Evans and Sutherland PS330 of the Laboratoire de Cristallographie Biologique (Strasbourg). We thank Dr. D. Moras for making it available and for his constant interest. This work was supported by the Centre de la Recherche Scientifique (CNRS).

REFERENCES

Amerein, B. (1981) PhD. Thesis, Université Louis Pasteur, Strasbourg.
Christiansen, J., Douthwaite, S.R., Christensen, A. and Garrett, R.A. (1985) EMBO J. <u>4</u>, 1019-1024.
Ehresmann, C., Baudin, F., Mougel, M., Romby, P., Ebel, J.P. and

Ehresmann, B. (1987) Nucleic Acids Res. <u>15</u>, 9109-9128.
Groebe, D.R. and Uhlenbeck, O.C.(1988) Nucleic Acids Res. <u>16</u>, 11725-11735.
Holbrook, S.R. and Kim, S.H. (1983) Biopolymers <u>22</u>, 1145-1166.
Jones, T.A. (1978) J. Appl. Cryst. <u>11</u>, 268-278.
Lavery, R. and Pullman, A. (1984) Biophys. Chem. <u>19</u>, 171-181.
Pieler, T. and Erdmann, V.A. (1982) Proc. Natl. Acad. Sci. U.S.A. <u>79</u>, 4599-4603.
Richmond, T. J. (1984) J. Mol. Biol. 178, 63-68.
Romaniuk, P.J. (1989) Biochemistry <u>28</u>, 1388-1395.
Romaniuk, P.J., Leal de Stevenson, I. and Wong, A.H. (1987) Nucleic Acids Res. <u>15</u>, 2737-2755.
Romaniuk, P.J., Leal de Stevenson, I., Ehresmann, C., Romby, P. and Ehresmann, B. (1988) Nucleic Acids Res. <u>16</u>, 2295-2312.
Romby, P., Westhof, E., Toukifimpa, R., Mache, R., Ebel, J.P., Ehresmann, C. and Ehresmann, B. (1988) Biochemistry <u>27</u>, 4721-4730.
Toots, I., Misselwitz, R., Böhm, S., Welfe, H., Villems, R. and Saarma, M. (1982) Nucleic Acids Res. <u>10</u>, 3381-3389.
Weiner, P.K. and Kollman, P.A. (1981) J. Comp. Chem. 2, 287-293.
Westhof, E., Dumas, P. and Moras, D. (1985) J. Mol. Biol. <u>184</u>, 119-145.
Westhof, E. (1988) In "Computer Aided Molecular Design" (I.B.C. Technical Services Ldt., London).
Westhof, E., Romby, P., Romaniuk, P., Ebel, J.P., Ehresmann, C. and Ehresmann, B. (1989) J. Mol. Biol. <u>207</u>, 417-431.
Westhof, E., Romby, P., Ehresmann, C. and Ehresmann, B. (1989) In Theoritical Biochemistry and Molecular Biophysics, (Beveridge, D.L. and Lavery, R.), Adenine Press, in press.

Water and Ions in Biomolecular Systems
Advances in Life Sciences
© 1990 Birkhäuser Verlag Basel

IONIC EFFECTS ON DNA ELECTROSTATIC AND ELASTIC STABILITY

M. O. Fenley, G. S. Manning, and W. K. Olson

Department of Chemistry, Rutgers University, New Brunswick, New Jersey 08903, USA

SUMMARY: We analyze the ionic dependence of the electrostatic and elastic stability of DNA within the framework of the counterion condensation theory. The ionic free energy difference between the B and Z conformational forms suggests a B->Z transition for short DNA (< 30 base pairs) at high salt and a Z->B->Z transition for polymeric DNA (Z->B with increasing salt at low salt and B->Z at high salt). The toroidal tertiary shape of superhelical DNA is found to be electrostatically more stable than the interwound form at low ionic strength. We define null DNA as a hypothetical form lacking phosphate charge and conclude by considerations of elastic instability that null DNA is curved (regardless of sequence).

INTRODUCTION

The polyelectrolyte character of DNA in solution, which arises from the presence of the charged phosphate groups in the sugar-phosphate backbone, affects both its electrostatic and elastic stability. In this short review we discuss the dependence of these DNA properties on the ionic environment (i.e., ion type and concentration) within the framework of the counterion condensation (CC) theory. Our approach in this paper is descriptive; the reader is referred elsewhere

(Manning, 1978, 1984, 1989b; Fenley et al., 1989) for more details of the theory, as well as for a complete analysis of our results and its comparison with experimental data.

COUNTERION CONDENSATION THEORY

IDEALIZED DNA-COUNTERION-SOLVENT MODEL

Within this model the charged groups of the polyion are represented as points on an infinite line with uniform spacing b between them. For double-helical DNA, b is the distance between phosphate projections on the helical axis. The only information concerning the polyion structure comes through this parameter. Counterions are treated as point charges. The solvent is regarded as a dielectric continuum with dielectric constant ε. The linear charge density parameter is $\xi = q^2/\varepsilon kTb$, where q is the protonic charge, k the Boltzmann constant, and T the absolute temperature.

For polyions with high charge density such that $\xi > N^{-1}$, a portion of counterions of valence N, sufficient to lower the net value of ξ to N^{-1}, condense on the polyion. This phenomenon is known as counterion condensation. According to the CC theory, for a polyion in the presence of a single counterion type, the equilibrium counterion binding fraction $N\theta_N$ (fraction of polyion charge neutralized by condensed counterions), in the limit of infinite dilution, is given by $1-(N\xi)^{-1}$. This prediction of $N\theta_N$ has been confirmed by numerous experimental studies. $N\theta_N$ is only dependent on the axial charge spacing of the polyion and the valence of the counterion. This model is expected to be accurate at low ionic strength, and experiments confirm its usefulness at higher salt as well. It is here used for the study of the elastic stability of DNA. As discussed in this paper, the idealized model is not adequate in all cases and thus, we introduce a more detailed model.

MORE DETAILED DNA-COUNTERION-SOLVENT MODEL

The DNA is treated in terms of its three-dimensional structure with the locations of all phosphate charges given by the coordinates of the phosphorus atoms. The size of the counterion is accounted for partially. The finite DNA length, which is of special interest when addressing oligomeric DNA, is considered. Although dielectric discontinuity and saturation effects have been considered, we restrict ourselves in this paper to treating the solvent as a dielectric continuum.

Within the CC framework, the use of this model provides an equilibrium counterion binding fraction which is dependent on counterion valence and "size", DNA length, and detailed geometry of phosphate charge distribution. The counterion binding fraction increases with DNA length and reaches a constant value for a sufficiently long DNA. The approach to the polymer limit is dependent on ionic strength (both ion charge and concentration). This prediction of a reduced binding fraction for oligomeric DNA is in accord with other studies (Olmsted et al., 1989). Thus, for oligomeric DNA the chain length is an important factor in determining the counterion binding fraction and thermodynamic properties derived from it. The three-dimensional structural details of the phosphate charge distribution affects the counterion binding fraction. We predict that θ(Z-DNA) is slightly greater than θ(B-DNA) while, the idealized model leads to the opposite trend. The binding fraction of the interwound and toroidal forms of superhelical DNA is greater than that of linear DNA with the same number of base pairs (bp). Also, we predict a larger binding fraction for the interwound form relative to the toroidal form, which is due to the more unfavorable phosphate-phosphate interactions in the former DNA structure. Thus, these results suggest that the binding fraction is not solely determined by ξ but also depends on the discrete phosphate charge distribution. We use this model to study the electrostatic stability of DNA.

METHODOLOGY

The CC theory provides the ionic free energy G for the DNA-counterion-solvent system. It is a sum of the electrostatic and mixing free energies. One electrostatic contribution to the ionic free energy originates from the interaction between the net phosphate charges (i.e., charges reduced by counterion condensation), in the Debye-Hückel approximation. This term is evaluated numerically when the more detailed DNA model is adopted, whereas it can be cast in an analytical form by using the idealized DNA model. The other electrostatic contribution to the ionic free energy, which was introduced to account partially for the counterion size, arises from the interaction of each net phosphate charge with its Debye-Hückel atmosphere. The mixing contributions to the ionic free energy arise from the mixing of free counterions, bound counterions, and solvent molecules. The value of the condensed counterion volume (v) (in the presence of added salt) cannot be uniquely determined by the theory. For the detailed model a numerical estimate of this parameter was made using molecular volume methods. We make the following assumptions: excess added salt (salt concentration much greater than phosphate concentration), same solvation state for the bound and free counterions, and territorial (delocalized) mode of counterion binding. From a knowledge of the DNA structure and condensed counterion volume the equilibrium counterion binding fraction is obtained by minimizing the ionic free energy (terms dependent on the binding fraction) with respect to the binding fraction. With the known values of θ_N and v, we can calculate the ionic free energy of DNA and thermodynamic properties derived from it (e.g., the work of stretching the DNA by phosphate-phosphate repulsion). One follows these steps independent of the model adopted, but using the appropriate free energy expressions which depend on the model and the assumptions made. A full description of the methodology is given elsewhere (Manning, 1984).

ELECTROSTATIC STABILITY OF DNA

As for any polyelectrolyte, the conformation of DNA in solution depends strongly on the ionic environment (e.g., ion type and concentration). At the secondary level of DNA structure, it is well established that the helix-coil transition (i.e., melting temperature) is dependent on ionic strength. Also, in the past fifteen years numerous experimental studies (for a review see, Jovin et al., 1987) on certain alternating purine-pyrimidine sequences showed that a right- to left-handed helical transition of DNA is induced by changes in ionic strength. Furthermore, at the tertiary level of DNA structure, there is experimental evidence (Gray, 1967; Bötlger & Kuhn, 1971; Vollenweider et al., 1976; Goulet et al., 1988) indicating that the relative stability of the interwound and toroidal forms of superhelical DNA is sensitive to changes in salt concentration. An understanding of the role of ionic strength in the conformational stability and transitions of DNA is of paramount biological importance. By using the idealized model, the CC theory has given an accurate analysis of the ionic free energy change involved in the DNA helix-coil transition. It also provides a qualitatively correct description of the salt dependence of the relative electrostatic stability of A- and B-DNA. According to the idealized model, raising the salt concentration stabilizes the right-handed B structure relative to the left-handed Z structure (both duplex forms have nearly equal linear charge densities), which is not in agreement with experiment. For a proper treatment of the salt-induced B-Z transition the idealized model is inadequate. The idealized model is not able to handle tertiary forms of DNA. Thus, these and other factors lead us to adopt the more detailed model. In this review, we address the B-Z and toroidal-interwound relative electrostatic stabilities as a function of ionic strength. The salt-induced B-Z transition has already been addressed by other theoretical approaches (Soumpasis, 1984; Hirata & Levy, 1989), which, however, are limited and leave open questions. Our goal is to examine the low salt regime, where controversies remain, and the influence of electrostatic end effects on this DNA conformational transition. Salt effects on superhelical DNA have not been treated previously by polyelectrolyte theory.

B-Z TRANSITION

For the B and Z (experimentally detected conformers: Z_I and Z_{II}) structures of DNA, we used idealized phosphorus coordinates generated from X-ray studies. Calculations of the ionic free energy difference between B- and Z- (Z_I and Z_{II}) DNA in 1:1 and 2:1 aqueous salt solutions were performed over a wide range of salt concentration (5×10^{-4} - 4 M) and DNA length (12 - 600 bp).

For oligomeric DNA (< 30 bp) in 1:1 salt solution, the B structure is electrostatically more stable than either of the Z structures over the salt range 5×10^{-4} - 1.4 M, after which the Z_{II} structure becomes electrostatically more stable relative to B-DNA. The critical monovalent salt concentration (i.e., 1.4 M) for the B-Z transition is in reasonable agreement with the experimental observation of 2.3 M (Pohl,

1983). The monotonic decrease of the ionic free energy difference between B- and Z-DNA (i.e., $\Delta G = G(Z\text{-}DNA) - G(B\text{-}DNA)$) is observed for both Z_I and Z_{II} structures, but the Z_I structure is never actually stabilized electrostatically with respect to B-DNA, although the difference in electrostatic free energy between the two forms approaches zero.

For polymeric DNA we predict that the electrostatic stability of the B structure drops from its maximum value (at ~ 0.1 M, in accord with experiment (Pohl, 1987)) by either increasing or decreasing monovalent salt concentration. Instead of a single B-Z_{II} transition at high salt, as in the case of oligomeric DNA, a double Z_{II}-B-Z_{II} transition is found for polymeric DNA free of electrostatic end effects (DNA length much greater than the Debye length). The contrast in behavior between ΔG for oligomeric versus polymeric DNA is due to the presence of end effects in the low salt regime for the oligomers. The critical low and high monovalent salt concentrations, for the polymeric DNA, occur at 2×10^{-3} M and 1.46 M, respectively. As expected, we find that the critical high salt concentration is virtually independent of DNA length (in agreement with experiment (Pohl, 1983)), since end effects do not play any role in this salt region. The Z_I structure exhibits the same non-monotonic behavior of ΔG, but again, as in the case of oligomeric DNA, the B structure is electrostatically more stable than Z_I-DNA, over the whole salt concentration range studied (but barely so at high salt).

The effect of a divalent counterion is to shift the critical salt concentrations at the B-Z_{II} transition point, for polymeric DNA, to smaller values, 0.008 M and 0.49 M, respectively, at low and high salt. The experimental value at high salt is 0.66 M (Pohl & Jovin, 1972).

Our calculations show that the electrostatic stability of the Z_{II} structure is greater than Z_I. The more favorable phosphate-phosphate interactions of Z_{II} cause this behavior.

Molecular mechanics (Kollman et al., 1982) and other theoretical studies (Klein & Pack, 1983; Broch et al., 1989), as well as experimental data (Behe, 1986) in the low salt regime, also indicate a greater stability of Z_{II}- over Z_I-DNA. Some experimental studies suggest that Z_I is more stable than Z_{II} at high salt. Molecular model building techniques and fiber diffraction studies (Sasisekharan et al., 1981) have shown that the transition between the Z_I and Z_{II} structures has little or no energy barrier. Our calculations are consistent with these observations, as they show only a small electrostatic free energy difference between the two Z forms. Thus, solvent (hydration) effects (which we have not attempted to analyze) may alter the equilibrium between Z_I- and Z_{II}-DNA, and therefore, perhaps, stabilize Z_I-DNA.

The greater electrostatic stability of polymeric Z- relative to B-DNA at very low ionic strength is a rigorous result in accord with the idealized model prediction, which is exact in the low salt limit. As expected, the use of the more detailed model is irrelevant in this limit. Since hydration (solvent) effects are expected to be minimal in comparison with electrostatic effects, in the low salt regime, we expect Z-DNA at low salt. Low salt B-Z transitions have been reported (Latha & Brahmachari, 1985) for the methylated and ethylated polymers poly(dG-methyl^{5}dC) and poly(dG-ethyl^{5}dC). Some researchers claim that the stabilization of Z-DNA at such low salt concentrations is due to extraneous Mg^{+2} present in the samples, but a recent study (Vorlíchová & Sági, private communication) shows that this is not the case for poly(dG-ethyl^{5}dC) and possibly for poly(dG-methyl^{5}dC) as well. Further experimental work is needed in order to clarify these conflicting results.

A comparison of the experimental value of the B/Z total free energy difference (0.44 kcal/mol-bp at 0.05 M NaCl) with the calculated value of the electrostatic free energy difference ($\Delta G = 0.20$ kcal/mol-bp for Z_{II} and 0.34 kcal/mol-bp for Z_I at 0.05 M) indicates that electrostatic effects contribute significantly to the experimentally observed value. But, the dependence of the B-Z equilibrium on base sequence, nature of coion and organic solvent indicate that solvent-mediated forces must play a role. For example, a recent study (Kagawa et al., 1989) shows that hydration effects may control the sequence dependence of the Z-DNA stabilization. Another study (McDonnell & Preisler, 1989) indicates that the differing ability of chemically related anions (with the same cation present) to drive the B-Z transition is probably due to their effects on the water structure surrounding DNA. For a quantitative description of the B-Z transition both ion and solvent effects should be taken into account. To date, theoretical studies have dealt only with one or the other effect, since it is difficult to treat both in an adequate manner. But, these studies have shed light on the factors that control the B-Z transition energetics. With further development and incorporation of appropriate statistical mechanical methods in computer simulation studies, a better understanding of DNA conformational stability will be achieved.

INTERWOUND VERSUS TOROIDAL DNA

Since the atomic coordinates for toroidal and interwound forms of superhelical DNA are not available from X-ray studies, it is necessary to obtain them using molecular modeling techniques. To generate the coordinates of the phosphorus atoms of these DNA forms, we used a differential geometry procedure (Olson & Cicariello, 1987) where the three-dimensional model is based on the deformation of B-DNA around preset interwound and toroidal space curves. In naturally occuring DNA there is roughly one superhelical turn for every twenty turns of the local B-helix (one superhelical turn per 200 bp), and the DNA supercoiling is almost exclusively left-handed. Based on this experimental evidence we chose the simplest model possible; both toroidal and interwound forms have one superhelical turn and total length of 200 bp. Each of the 200 bp structures is characterized by a fixed linking number of 20. We calculate the ionic free energy difference between the interwound and toroidal forms (ΔG = G(interwound) - G(toroidal)) as a function of monovalent salt concentration.

ΔG decreases monotonically with increasing salt concentration and changes sign ($\Delta G < 0$) at 0.02 M. These results indicate that the toroidal form is favored electrostatically over the interwound form at low ionic strengths. This behavior is expected, since the phosphate groups are, on average, more closely spaced in the compact interwound form. Thus, our results show that the relative electrostatic stability of the toroidal and interwound forms with the same linking number is highly sensitive to changes in ionic strength. These results are in accord with a previous study (Olson & Cicariello, 1987) where an analysis of the salt dependence of the Debye-Hückel energy for these same forms was made. Hydrodynamic studies (Gray, 1967; Bötlger, 1971) indicate that the decrease in the sedimentation coefficient below 0.01 M is due to the stabilization of the more expanded toroidal form of DNA. These experimental studies as well as an electron microscopy study (Vollenweider et al., 1976) suggest that a salt-induced interwound->toroidal transition occurs at low salt. Thus, the electrostatic energy seems to be an important factor in determining the tertiary DNA structure stability (transition). On the other hand, computer simulation (Hao & Olson, 1989) and other theoretical studies based on an elastic DNA model, with no electrostatic energy included, predict that the interwound form is favored over the toroidal form. A theoretical study based on both elastic and electrostatic DNA model is warranted for a better understanding of the dominant factors which contribute to the superhelical DNA energetics.

DNA ELASTIC STABILITY

The DNA polymer is unusually stiff (resilient) on a scale of 100 bp. We have been interested in the possibility of modeling DNA on this scale as a uniform, isotropically bending rod, amenable to classical mechanical analysis. This point of view is not new. It lies at the basis of a familiar theory of polymer persistence length (Landau & Lifshitz, 1958) and has been throughly exploited to investigate the twisting and writhing of closed circular superhelical DNA (Hao, 1988). That on the scale of a few bp DNA bends anisotropically into a preferred groove is not important for this type of modeling, which uses a coarse-grained Hooke's Law constant for bending on a scale of many double-helical turns, with the groove appearing now on this side, now on that, effectively generating an isotropic bend, smooth on the scale considered.

We have been able to find some interesting relations that enrich the general theory of polymer persistence (Manning, 1986a, 1986b), provide insight into the possible behavior of stressed DNA in chromatin (Manning, 1987, 1988), and indicate the role of the phosphate charge of DNA in maintaining a straight double-helical axis (Manning, 1989a, 1989b). The concept of elastic instability is accorded a central role. We choose one paper from this series for review (Manning, 1989b).

The phosphate charge on the DNA outer surface holds the polymer in a state of tensile stress, like a stretched spring. This aspect of the structure of DNA becomes evident when ionic strength is increased, thereby screening the phosphate-phosphate repulsions more effectively and weakening the stretching force. Then, like a stretched spring responding to diminishing tension, the contour length of DNA is observed to become shorter. In fact the measured contraction agrees with a quantitative relation that we had previously derived, namely, that the decrease of contour length should be linearly correlated with the logarithm of salt concentration.

According to the principle of counterion condensation, supported by a wealth of data, the extent of phosphate charge neutralization does not significantly increase on augmentation of the ionic strength, at least not in the range of unexceptional concentration. Instead, the fraction of counterions bound to DNA does not change much, yielding, for example, in the monovalent case, about 76% neutralization of the phosphate charge. Titration with multivalent cations, however, can increase the degree of neutralization up to 94% in the case of spermine, a tetravalent oligoamine.

The observation of contour length contraction when simple monovalent salt is added is replaced by a more dramatic event when over 90% of the phosphate charge is annihilated by bound multivalent cations. The long DNA polymer collapses to a viral-sized particle, with the double-helical axis wound circumferentially into the configuration of a stowed garden hose (rod-shaped particles are also observed). Since this phenomenon does not appear to depend on the specific characteristics of the phosphate-neutralizing agent, the question arises of the properties of null DNA, that is, of DNA deprived of its anionic nature.

We have demonstrated that the double-helical axis of null DNA is probably buckled, thus explaining the high degree of ordered curvature in DNA packaged by extensive charge neutralization. The situation must be distinguished from that occurring with certain nonrandom sequences of DNA, which are curved when ionized due to structural peculiarities specific to the unusual sequence. We have in hand in our discussion a curvature induced in any (given the variety of DNAs for which the packaging occurs) DNA sequence on the scale of 100 bp when the phosphate charge is neutralized. We suspect the implication of solvent mediated forces, but our analysis is restricted to the demonstration that buckling occurs for any DNA sequence with nonexceptional persistence length and does not extend to the analysis of the molecular-level forces responsible for it. We conclude that on the scale of a persistence length null DNA is like a curved relaxed spring that can be stretched out straight only by applying a tension force on it (phosphate-phosphate repulsion on removal of the charge-neutralizing multivalent cation). In other words the 100 bp scale on which the double-helical axis of DNA is observed to be straight under ordinary ionic conditions is supported by the phosphate charge. On removal of the charge, the "natural" shape of the DNA molecule (now null DNA) emerges as curved. (The curvature can be concentrated at a "point", thus explaining the high density of kinks in packaged forms that are rod-shaped instead of gradually curved).

The line of argument used to reach this conclusion is rooted in Eulers's theory for buckling of a resilient rod under compression along the long axis. Euler derived a formula indicating that when the compressive force is less than a certain critical value (depending on the length of the rod and the Hookes's Law bending constant), then the rod responds to the applied force with a simple contraction along its axis. But if the force exceeds the critical value, the straight axis becomes elastically unstable and yields to a curved form. The rod buckles.

In the application to DNA, the compressive force is the restoring force in the DNA "spring" that balances the phosphate-phosphate stretching force in the stable structure of DNA in its usual ionization state (Every interatomic force in a static structure must be balanced by another; otherwise, the structure cannot be static). On the molecular level the restoring force may be hard to describe. It consists of the net resultant of all forces generated by structural distortion caused by the phosphate-phosphate interactions. On the level of Eulers's theorem, however, we need to know only that the restoring force is numerically equal to (and oppositely directed from) the phosphate-phosphate resultant tension. The latter can be estimated from polyelectrolyte theory.

Consider, now, DNA "at the instant" of charge neutralization. The phosphate tension has been annihilated, but the restoring force still operates. The structure, thus subjected to an unbalanced compression (the restoring force that used to be, but is no longer, balanced by the phosphate stretch), must change. If the restoring compression is less than Euler's critical force, the structure merely contracts and the final static structure of null DNA would be a slightly contracted form of ordinary DNA. But if the compression is in excess of critical, then null DNA buckles. Its static shape is curved, or possibly, kinked.

From Euler's formula and other considerations we have derived a buckling condition, which, if fulfilled, indicates that null DNA is indeeed buckled. In essence, it states that, for buckling, the restoring compression must be greater than the Euler critical force:

$$F^3 > 4\lambda\,(kT)^3\,/\,\pi^2\,R^4$$

The formula contains only quantities characteristic of DNA in its ordinary state of ionization. F is the phosphate tension force, derivable from polyelectrolyte theory. The length λ is the persistence length of DNA ($\sim$ 150 bp). R is the radius of the elastic rod that best models DNA. (Since the elastic energy includes an energy originating in the ionic interactions among the phosphates and the Debye atmosphere of small ions, R is not necessarily derivable from crystallographic data. It has been estimated at 4.4 nm in 0.01 M salt, for example, compared with about 1.0 nm for the crystallographic radius of DNA and about 3.0 nm for the Debye length at this ionic strength). Insertion of the numbers suggests that the buckling condition is easily satisfied in the DNA/null DNA system, and our conclusion about the local curvature of DNA with neutralized phosphate groups follows.

Acknowledgements: This work was supported by USPHS grants GM20861 (GSM) and GM34809 (WKO). One of us (M.O.F.) would like to express her gratitude to Rutgers University for fellowship support during the past two years. Calculations were performed at the Rutgers Center for Computational Chemistry.

REFERENCES

Behe, M. J. (1986) Biopolymers $\underline{25}$, 519-523.
Bötlger, M., and Kuhn, W. (1971) Biochim. Biophys. Acta $\underline{254}$, 407-411.
Broch, H., Viani, R., Grassi, H., and Vasilescu, D. (1989) Int. J. Quantum Chem., in press.
Fenley, M. O., Manning, G. S., and Olson, W. K. (1989), submitted.
Goulet, I., Zivanovic, Y., Prunell, A., Revet, B. (1988) J. Mol. Biol. $\underline{200}$, 253-266.
Gray, H. B.,Jr. (1967) Biopolymers $\underline{5}$, 1009-1019.
Hao, M.-H. (1988) Ph.D. Thesis, Rutgers University, New Brunswick, New Jersey, USA.
Hao, M.-H., and Olson, W. K. (1989) Macromolecules $\underline{22}$, 3292-3303.
Hirata, F., and Levy, R. M. (1989) J. Phys. Chem. $\underline{93}$, 479-484.
Jovin, T. M., Soumpasis, D. M., and McIntosh, L. P. (1987) Annu. Rev. Phys. Chem. $\underline{38}$, 521-560.
Kagawa, T. F., Stoddard, D., Zhou, G., and Ho, P. S. (1989) Biochemistry $\underline{28}$, 6642-6651.

Klein, B. J., and Pack, G. R. (1983) J. Biol. Phys. 11, 23-25.

Kollman, P. A., Weiner, P. K., Quigley G., and Wang, A. H.-J. (1982) Biopolymers 21, 1945-1969.

Landau, L. D., and Lifshitz, E. M. (1958) In: Statistical Physics, Addison-Wesley, Reading, Massachusetts, USA.

Latha, P. K., and Brahmachari, S. K. (1985) J. Scientific Ind. Res. 45, 521-533.

Manning, G. S. (1978) Quart. Rev. Biophys. 11, 179-246.

Manning, G. S. (1984) J. Phys. Chem. 88, 6654-6661.

Manning, G. S. (1986a) Phys. Rev. A 34, 668-670.

Manning, G. S. (1986b) Phys. Rev. A 34, 4467-4468.

Manning, G. S. (1987) Quart. Appl. Math. 45, 809-815.

Manning, G. S. (1988) Phys. Rev. A, 38, 3073-3081.

Manning, G. S. (1989a) J. Biomol. Struct. & Dyn. 6, 877-889.

Manning, G. S. (1989b) J. Biomol. Struct. & Dyn. 7, 41-61.

McDonnell, N. B., and Preisler, R. S. (1989) In: Book of Abstracts:Sixth Conversation in Biomolecular Streodynamics, (R. H. Sarma, Ed.), Albany, New York, pp. 222.

Olmsted, M., Anderson, C. F., and Record, M. T., Jr. (1989) Proc. Natl. Acad. Sci. USA, in press

Olson, W. K., and Cicariello, J. (1987) Ann. N. Y. Acad. Sci. 482, 69-81.

Pohl, F. M., and Jovin, T. M. (1972) J. Mol. Biol 67, 375-396.

Pohl, F. M. (1983) Cold Spring Harbor Symp. Quant. Biol. 48, 113-117.

Pohl, F. M. (1987) In: Structure, Dynamics and Function of Biomolecules, (A. Ehrenberg, A. Rigler, A. Graslund, and L. Nilsson, Eds.), Springer-Verlag, Heidelberg, pp. 224-228.

Sasisekharan, V., Bansal, M., Brahmachari, S. K., and Gupta, G. (1981) In: Biomolecular Stereodynamics, vol. 1, (R. H.Sarma, Ed.), Adenine Press, New York, pp. 123-149.

Soumpasis, D. M. (1984) Proc. Natl. Acad. Sci. USA 81, 5116-5120.

Vollenweider, H. J., Koller, TH., Parello, J., and Sogo, J. M. (1976) Proc. Natl. Acad. Sci. USA 73, 4125-4129.

Water and Ions in Biomolecular Systems
Advances in Life Sciences
© 1990 Birkhäuser Verlag Basel

COPPER IONS AT TRACE LEVEL AND DNA FUNCTIONING : STUDIES PERFORMED WITH ^{64}Cu AND ^{67}.Cu

S. Apelgot and E.Guillé

Institut Curie, Section de Physique et Chimie, 11 rue P. et M. Curie, 75231 Paris Cedex 05, France, and Biologie Moléculaire Végétale (UA 1128) Université Paris Sud, 91405 Orsay, France

SUMMARY: The transmutation of radioactive copper atoms incorporated at a late time inside the DNA of mammalian cells leads to cell death. The lethal event is irreparable and the lethal efficiency is high. In order to explain these results, these copper atoms should occupy a strategic position and play a fundamental role. This role was connected to that of metal traces in the software of a microcomputer, and a DNA organisation in 3 units was suggested.

It is well known that different metals at trace concentration are essential for life. One of their most important roles is as a cofactor of enzymes. The discovery of metal ions bound not randomly along the DNA molecules opened new fields for the understanding of the various roles of metals for life. From neutron activation analysis, Cr^{3+}, Sb^{2+}, Fe^{2+}, Zn^{2+} and Co^{2+} were found in DNA extracted from lymphocytes (Andronikashvili et al., 1974) and their content was related to the tissues studies (Andronikashvili et al., 1976). Cd^{2+}, Pb^{2+} and Cu^{2+} were found in reiterative DNA sequences of DNA molecules extracted from various organisms (bacteria, plants, animals) (Sissoëff et al., 1976). It was always

46

established that these metal traces are endogeneous and not artifacts due to the various steps of the extraction procedures
(Sissoëff et al., 1976 ; Guillé et al., 1981).

Although the presence of metal ion traces are clearly evidenced
in all DNA studied, their specific role was not clearly understood. It seemed possible to shed light on this role by using
radioactive metal isotopes, since the transmutation of a radioactive isotope can lead to a lethal effect when this transmutation
occurs inside the DNA of a cell (Hershey et al., 1951). Beginning
in 1951, many experiments have been performed in this field with
constitutive DNA atoms (^{14}C, ^{32}P, ^{33}P) (for a review see Apelgot,
1983). By using two different radioactive isotopes of copper,
^{64}Cu and ^{67}Cu, and two different mammalian cell lines, it was
possible to establish that : 1) radioactive copper atoms remain
inside purified DNA and are not randomly distributed ; this distribution is different for the two cell lines studied (Grisvard
et al., 1989) ; 2) the transmutation of ^{64}Cu or ^{67}Cu atoms give
rise to a strong and irreversible inhibition of DNA synthesis
(Apelgot et al., 1989) ; 3) the transmutation of these isotopes
incorporated at a late time inside the DNA molecule of the mammalian cells studied leads to cell death and the lethal event is
irreparable (Apelgot et al., 1989).

These results underline the fundamental role played by some
copper atoms, those incorporated late inside the DNA molecules.
By connecting the role of these few copper atoms to that of metal
traces in the software of a microcomputer, DNA organisation in
three units inside the cells was postulated.

Recall of the lethal effect via transmutation (Apelgot et al.,
1989): During a decay process a radioactive atom most frequently
becomes a stable atom characterized by a different atomic number
leading to different chemical properties; this is the transmutation per se. At the same time one or many ionizing particles are
emitted. It is clear that both events (transmutation and ionizing
particles) are characterized by a probability P_t and P_i to kill
cells. Experiments performed since 1951 with many radioactive
isotopes and many different kinds of cells from phages to mammalian cells have shown with no exception that P_t (the lethal probability of a transmutation process) is always much lower than

P¡ (lethal probability of a ionizing process) **except when the transmutations have occurred inside the DNA molecules** of the cells.

REPORTS ON PREVIOUS RESULTS

Experiments were performed with two types of mammalian cells, A549, a human malignant cell line, and CV1, a simian cell line. ^{64}Cu and ^{67}Cu were used as chloride salt. Unlabelled $CuCl_2$ was always added in order to obtain a standard final copper concentration of 1 to 2 µg/ml.

<u>Biochemical results</u> (Grisvard et al., 1989): After purification of DNA extracted from A549 or CV1 cells labelled with ^{64}Cu or ^{67}Cu,

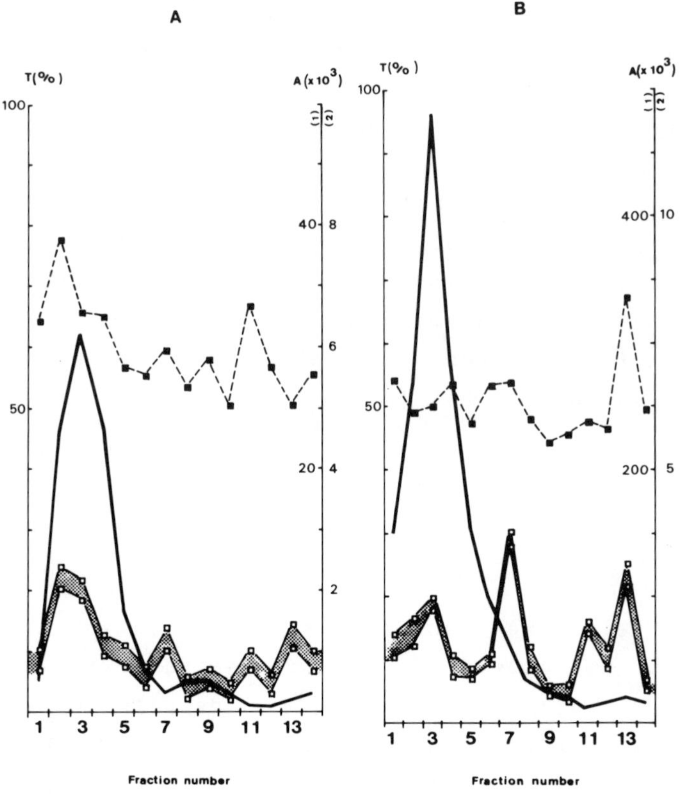

Figure 1.
CsCl density gradient of DNA from CV1 (A) and A549 (B) cells labelled with ^{67}Cu.
(—): transmission % at 253.7 nm recorded at the time of collecting the fractions. (■) and scale 1: dpm measured on each fraction before dialysis. (□) and scale 2: dpm measured on each fraction after dialysis; the 2 curves surround the experimental results taking into account the errors. The fractions were collected from the bottom of the tubes (A = dpm/ml).

radioactive copper atoms were recovered inside the DNA fractions (Fig. 1). The results obtained clearly show that some radioactive copper atoms can reach DNA and established with it bonds strong enough to hold on in spite of such energetic treatments as ethanol precipitation or centrifugation in CsCl gradient. This copper-DNA association is not an artifact established during the different manipulations as proved by control experiments where ^{64}Cu was added at different steps of the DNA extraction procedure (Apelgot et al., 1981). Analytical ultracentrifugation was performed on the fractions containing a high amount of DNA and of radioactivity. It was observed that these fractions were enriched in satellite DNA, localized generally in the constitutive heterochromatin areas. This was, in some cases, confirmed by a supplementary centrifugation in Ag^+-Cs_2SO_4 gradients.

The lethal effect (Apelgot et al., 1984 and 1989): ^{64}Cu decays give rise to a strong and irreversible inhibition of DNA synthesis. This inhibition occurs after a certain period of contact depending on the cell type (Fig. 2). Furthermore, the transmutation of ^{64}Cu or ^{67}Cu incorporated in the non-synchronized CV1 or A549 cells leads to a lethal effect characterized by an exponential survival curve (Fig. 3). The salient result was that these two radionuclides have identical and high lethal efficiencies despite their different disintegration schemes (Fig. 4). It was also observed that the lethal event occurs inside the DNA molecules of the mammalian cells studied and that this lethal event is irreparable.

DISCUSSION

For the two different cell lines studied, the transmutation of ^{64}Cu or ^{67}Cu inside the DNA molecules lead to a lethal effect characterized by an identical lethal efficiency, despite their different disintegration schemes. The new atom appearing consequent to their disintegration cannot be at the origin of the lethal effect, since this new atom is different for ^{64}Cu and ^{67}Cu decays (Fig. 4). Therefore, the lethal effect can only be the consequence of the decay phenomenon itself. For all radioactive atoms, the decay phenomenon takes place over a very short time (10^{-15} sec.) and corresponds to an abrupt and sudden modification of an unstable atom nucleus. These results focus on the fact that the decay of only a few radioactive copper atoms has a consequence as severe as the death of a mammalian cell. We have also observed a lethal effect as a consequence of the decay of radioactive zinc atoms (^{65}Zn) incorporated in mammalian cells (Joseph, 1984). What might the fundamental role of copper and zinc inside the DNA

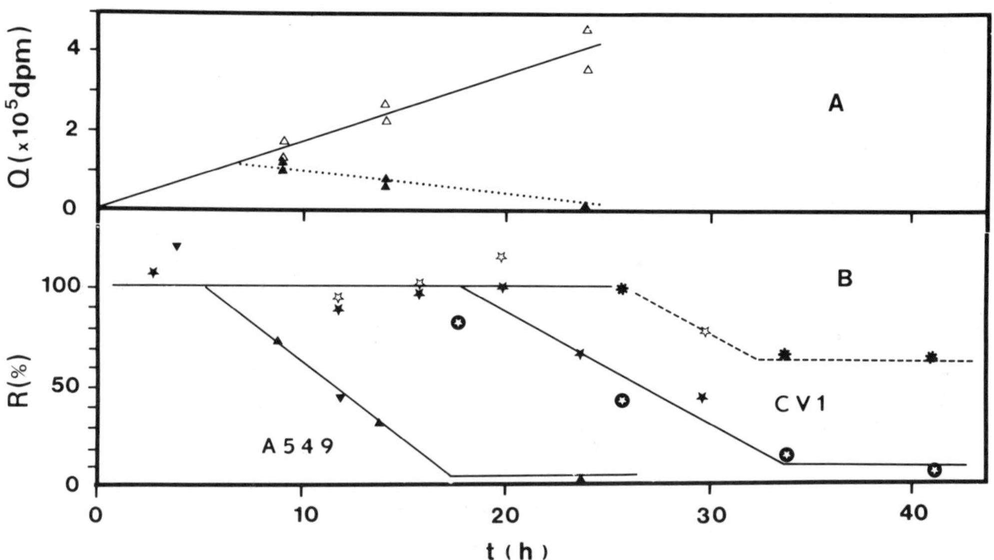

Figure 2. (^{3}H)thymidine incorporation into DNA of ^{64}Cu treated cells. A549 cells were grown for different times in the presence of 390-440µCi/ml (at t=0) of ^{64}Cu used as ^{64}CuCl$_2$ and CV1 cells in the presence of 200 or 400µCi/ml (at t=0) of ^{64}CuCl$_2$. After the indicated time intervals the medium was discarded and 1h pulse-labelling with (^{3}H)thymidine was performed. **A**: (^{3}H)thymidine incorporation (Q) into cell DNA for either control (open symbols) or labelled (solid symbols) A549 cells; B: Ratio (R) of (^{3}H)thymidine incorporation into ^{64}Cu treated cells relative to that of control cells. Two independent experiments were always performed: A549 cells and ^{64}CuCl$_2$ (▼, ▲) at about 400 Ci/ml (at t=0); CV1 cells and ^{64}CuCl$_2$ at about 400 Ci/ml (at t=0) (✶ ; ✪), or at about 200 Ci/ml (at t=0) (☆ ; ✳).

50

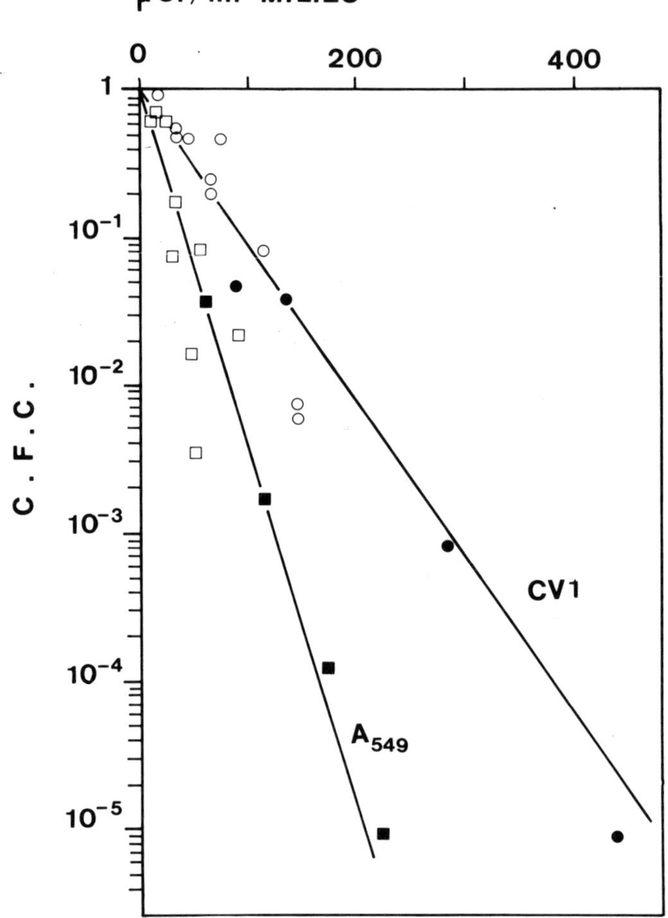

Figure 3. Colony-forming capability (CFC) of CV1 and A549 cells (asynchronous cultures) labelled with ^{64}Cu or ^{67}Cu used as $CuCl_2$: CFC is expressed as a function of the radioactivity present in the growth medium. ^{64}Cu results: clear symbols; ^{67}Cu results: dark symbols.

molecule be, since a local, but abrupt, disturbance in these metal atoms brings about the death of mammalian cells with a high efficiency ? One might think that a fundamental structure containing copper and/or zinc is implicated in the management of the overall behaviour of cells. Copper and zinc have natural access to this structure, probably in the chromatin areas of mammalian cells. Their concentration in this structure depends on cellular type (Tashima et al., 1981), as well as on environmental variations (Guillé et al., 1981).

It is well known that metals, at trace levels, are one of the fundamental parts of software of microcomputers. We would like to

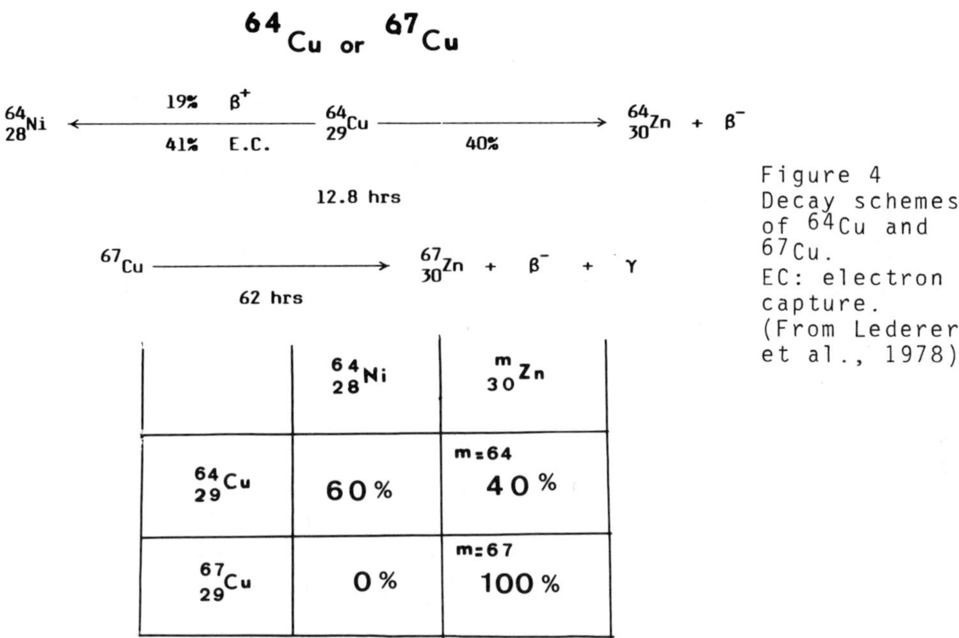

Figure 4
Decay schemes
of 64Cu and
67Cu.
EC: electron
capture.
(From Lederer
et al., 1978)

	64/28 Ni	m/30 Zn
64/29 Cu	60 %	m=64 40 %
67/29 Cu	0 %	m=67 100 %

suggested, as a working hypothesis, that metals, also at trace
levels, might be, as in the software of a microcomputer, one of
the key-points for cellular DNA functioning.

THE SUGGESTED MODEL

We tried to foresee a cell organization able to fulfil the basic
components of the software of a microcomputer (Apelgot and Guillé,
1988). As there are three esential functions in software: control,
exchange and processing units, cellular DNA should also be organized along the lines of three units. A small part of the constitutive and interstitial DNA sequences would be the "control unit",
The remaining constitutive DNA sequences localized at the periphe-
ry of the nucleus would be the major part of the "exchange unit",
detecting environmental variations (inside and outside the cell)
and dealing with them only with the agreegment of the "control

unit". The remaining interstitial heterochromatin sequences dise-
minated throughout the entire genome would be the "processing
unit", that is, the exchange unit effectors, answering the orders
sent to them by the teleaction process (Wells et al., 1983) from
the exchange unit. The interstitial heterochromatin sequences
must be localized in the vicinity of the genes in euchromatin
areas to subject them to their control.

These three units (control, exchange and processing) must be
permanently interconnected. Moreover, in order to be able to cope
rapidly with new cellular situations, their organization must not
be rigid. The "control unit" only, which maintains the entire
program for all cellular life, must be hyperprotected to prevent
destabilization of the cells. In some specific situations, such
as stress and cancer induction, this control unit will still
change in specific ways including topological modifications with
generally dramatic consequences for the future of the cell.
Furthermore, it is clear that a better defined control hierarchy
will be the result of a greater complexity of the organism. The
two other units (exchange and processing) are more flexible. A
few of their components can even be lost without dramatic consequences, as long as the original matrix or some copies remain in
the genome. Numerous facts support this point of view. It is well
known that translocation might be favoured between small reiterative DNA sequences when a mitotic crossing-over takes place.

Our model implies two considerations:

a) The metals bound to reiterative DNA sequences must not
remain at the same site during the life time of a cell but must
move and/or change according to physiological and/or pathological
states.

b) The DNA sequences of the constitutive heterochromatic areas
related to the control unit must be preserved with great accuracy;
conversely the constitutive and interstitial heterochromatic
areas related to the exchange and the processing units can be
preserved with less accuracy.

These two characteristics are indeed observed in all already
known biological processes.

<u>CONCLUSION</u>: Genetics deal with only a small percentage of total cellular DNA. It is difficult to accept that the rest of the cellular DNA plays no role. To take into account the fundamental role played by some copper atoms inside the DNA molecule as evidenced by the results reported, we hypothesized that cellular DNA should be organized in three units (control, exchange and processing) as is the software inside a microcomputer where metal traces are fundamental. The well known DNA genes would thus correspond to the processing unit and could function only if the other two units (control and exchange) are functional. The regulating DNA, which is beginning to be studied, would correspond to the exchange unit and the rest of the DNA where the metal atoms are most important would correspond to the control unit. Nothing is yet known about this strategic DNA portion.

REFERENCES

Andronikashvili, E.L., Mosulishvili, L.M., Belokobilski, A.J., Karaboudze, N.E., Tevzieva, J.H., and Efremova, E.V. (1974) Cancer Res. <u>34</u>, 271-274.

Andronikashvili, E.L., Belokobilski, A.J., Mosulishvili, L.M., Karabadze, N.E. and Shonya, N.I. (1976) Dokl. Akad. Nank. SSSR <u>227</u>, 1244-1252.

Apelgot, S., Coppey, J., Grisvard, J., Guillé, E. and Sissoëff, I. (1981) Cancer Res. <u>41</u>, 1502-1507.

Apelgot, S. (1983) Int. J. Radiat. Biol. <u>43</u>, 95-101.

Apelgot, S., Coppey, J., Grisvard, J., Guillé, E. and Sissoëff, I. (1984) C.R. Ac. Sci. Paris <u>298</u>, 31-34.

Apelgot, S. and Guillé, E. (1988) In: DNA Damage by Auger Emitters (K.F. Baverstock and D.E. Charlton Eds), Taylor and Francis, pp. 81-88.

Apelgot, S., Coppey, J., Gaudemer, A., Grisvard, J., Guillé, E., Sasaki, I. and Sissoëff, I. (1989) Int. J. Radiat. Biol. <u>55</u>, 365-384.

Grisvard, J., Guillé, E., Sasaki, I., Sissoëff, I., Valenza, A., Apelgot, S., Coppey, J. and Gaudemer, A. (1989) Biochemical Trace Element Research <u>20</u>, 207-217.

Guillé, E., Grisvard, J. and Sissoëff, I. (1981) In: Systemic Aspects of Biocompatibility (D.E. Williams Ed) CRC Press, USA, <u>1</u>, pp. 39-85.

Hershey, A.D., Kamen, M.D., Kennedy, J.W. and Gest, H. (1951) J. Gen. Physiol. <u>34</u>, 305-319.

Joseph, A. (1984) Contribution à l'étude de la distribution du
    zinc dans des tissus sains et tumoraux. Thèse (Paris, 22 Nov.)
Lederer, C.M., Hollander,Y.M., and Perlman, I. (1978) Table of
    Isotopes, John Wiley, Chichester.
Sissoëff, I., Grisvard, J. and Guillé, E. (1976) Progr. Biophys.
    Mol. Biol. 31, 165-199.
Tashima, M., Calabretta, B., Torelli, G., Scofield, M., Matzel, A.
    and Saunders, F. (1981) In: Proceedings of the National
    Academy of Sciences (USA) 78, 1508-1512.
Wells, R.D., Blakesley, R.W., Hardis, S.C., Horn, G.T., Larson,
    J.E., Seling, E., Bord, J.F., Chan, H.W., Jensen, K.F., Nes,
    J.F. and Wartell, R.M. (1977) Critical Reviews in Biochemistry
    4, 305-340.

Z-DNA TRANSCONFORMATION : A QUANTUM MOLECULAR APPROACH

H. BROCH, R. VIANI, H. GRASSI, D. VASİLESCU

Laboratoire de Biophysique. Parc Valrose 06034 NICE . France

SUMMARY: The polymorphism of nucleic acids is now well established and we know enough about the so-called Z-DNA form of left-handed helix to be able to undertake a quantum molecular approach of the various possible transconformations. This paper reports some of the first results we obtain by computations on our disugar-triphosphate model, essentially in the study of the varying conformations opened to the phosphate groups and the optimization of the forms named Z_I and Z_{II}.

OBJECT OF THE STUDY

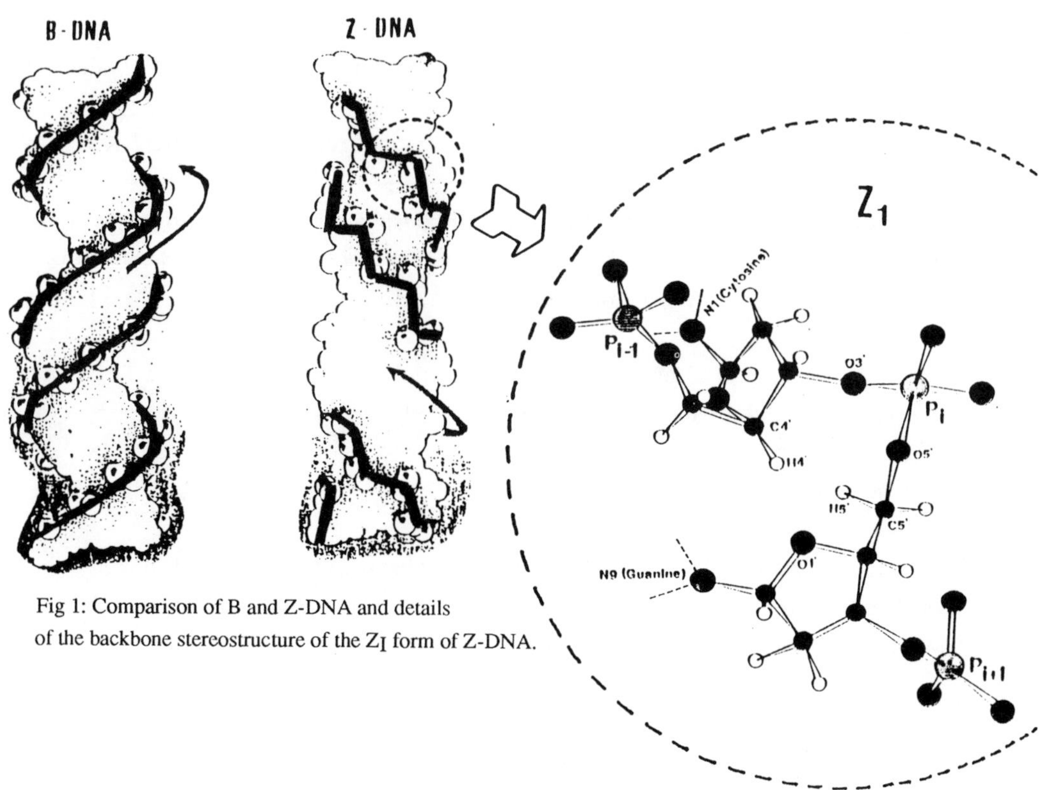

Fig 1: Comparison of B and Z-DNA and details of the backbone stereostructure of the Z_I form of Z-DNA.

The description of the principal stereostructures we can observe in nucleic acids (Arnott et al. 1986) clearly shows that the polymorphism of these molecules is a large field of investigation. X-ray diffraction analysis have shown that DNA can take a zig-zag conformation so named Z-DNA (cf Fig. 1).

The Z conformation has been discovered by Pohl and Jovin (1972) in poly (d(G-C)) high salt solutions and the first X-ray structure determinations were made by Wang et al. (1979, 1981), Drew et al. (1980) and Crawford et al. (1980). Now, a set of infrared bands characteristics of DNA conformation have been put in evidence to identify Z conformations (Taboury et al. 1985)

The major changes between the classical right-handed B-DNA and Z-DNA are that Z-DNA is a left-handed helix, necessitates sequence of alternating pyrimidines and purines (principally deoxycytidine and deoxyguanosine) and that deoxyguanosine has a special conformation. This special conformation is characterized by a syn conformation for the base (instead of the familiar anti of B-DNA) and by a C3' endo puckering for the deoxyribose ring (instead of the C2' endo of B-DNA).

Another important fact is that the sugar-phosphate backbone is kinked in one unit on two, by the change of the value of the exocyclic rotational parameter from its normal 60° B-DNA value to the 180° or 160° one.

Crystallographic data have also shown various conformations for the left-handed DNA, named Z_I, Z_{II} and Z' (for example of the Z_I form, cf Fig. 1).

On another hand, N2 amino group (base) and O3' phosphate group are bridged by a water molecule in Z-DNA, whereas in Z'-DNA the phosphate is repelled by a chloride bound to N2; moreover, in Z' structures, the sugar of guanosine is in a C1' exo puckering.

The object of the study is to undertake a quantum molecular approach of the DNA sugar-phosphate backbone conformations with the aim to answer to various questions as:
- is there a "family" of Z-DNAs?
- what are the stabilizing factors for these structures ?
- are Z_I and Z_{II} separated species ?
- what are the conformational paths between B and Z and between the various Z forms ?
- what are the correlated necessary changes in backbone for these transconformations ?
- what is the phosphate "motion" dependence versus the other conformational parameters ?...

May be it will be possible in a near future, after the determination of the allowed zones of all the rotational parameters for Z-DNAs, to study the "macroscopical" aspect in a geometrical way as we have already done for the "normal" DNA (Broch et al. 1977).

PROCEDURE

The model we used, with all the rotational parameters studied in this work, is the dideoxyribose-triphosphate model illustrated in Fig. 2.

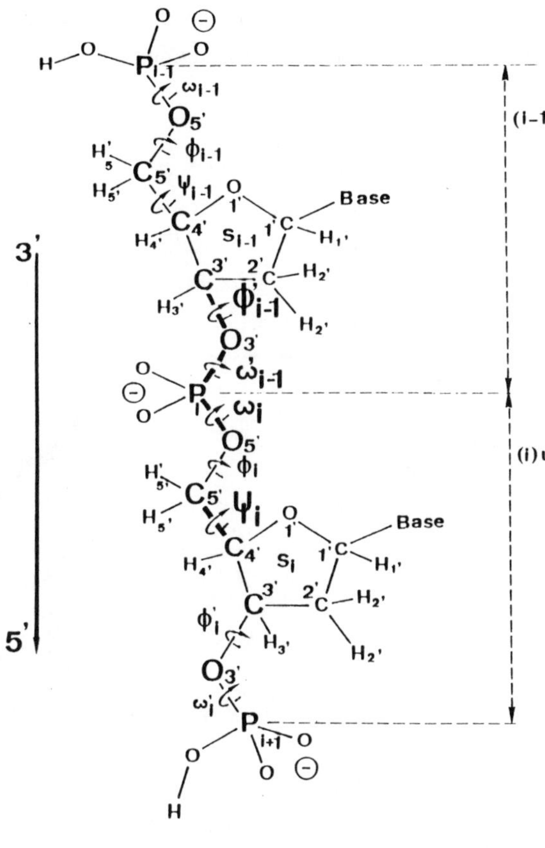

Fig. 2: the dideoxyribose-triphosphate model.

The procedure is to compute and minimize the energy of the molecule as a function of the various rotational parameters.

The computations were made with the help of PCILO and flexible simplex methods.

For details on all aspects of the procedure (methods, geometrical parameters, angular definitions, "classical" and "standard" values,...) see Broch and Vasilescu (1989) and Broch et al. (1989).

With regard to the 3'-->5' direction (i-1 unit to i unit) the sugar puckerings are noted ^{2}E for C2' endo and ^{3}E for C3' endo; for the rotation around the exocyclic bond (parameter ψ), we use gg and gt symbols for values about 60° and 180°.

58

RESULTS AND DISCUSSION

On the basis of crystallographic analyses, it can be infered that "left-handed double helical Z-DNA molecules can exist as a family of structures with varying conformations of phosphate groups in the sugar-phosphate backbone" (Wang et al. 1981). As emphasized by these authors, the backbone seems to be the principal part in the understanding of the various stereostructural possibilities for left-handed DNA.

So as first step, we have computed the conformational energetic maps $E(\omega'_{i-1}, \omega_i)$ with the other rotational parameters positioned on the various B and Z structure values, in the aim to understand the different transconformations allowed for the sugar-phosphate backbone.

We just want to present here an example of the obtained results. Fig. 3 represents the $E(\omega',\omega)$ energetic map obtained with the sugars in $^2E^3E$ and ψ_i in gt position.

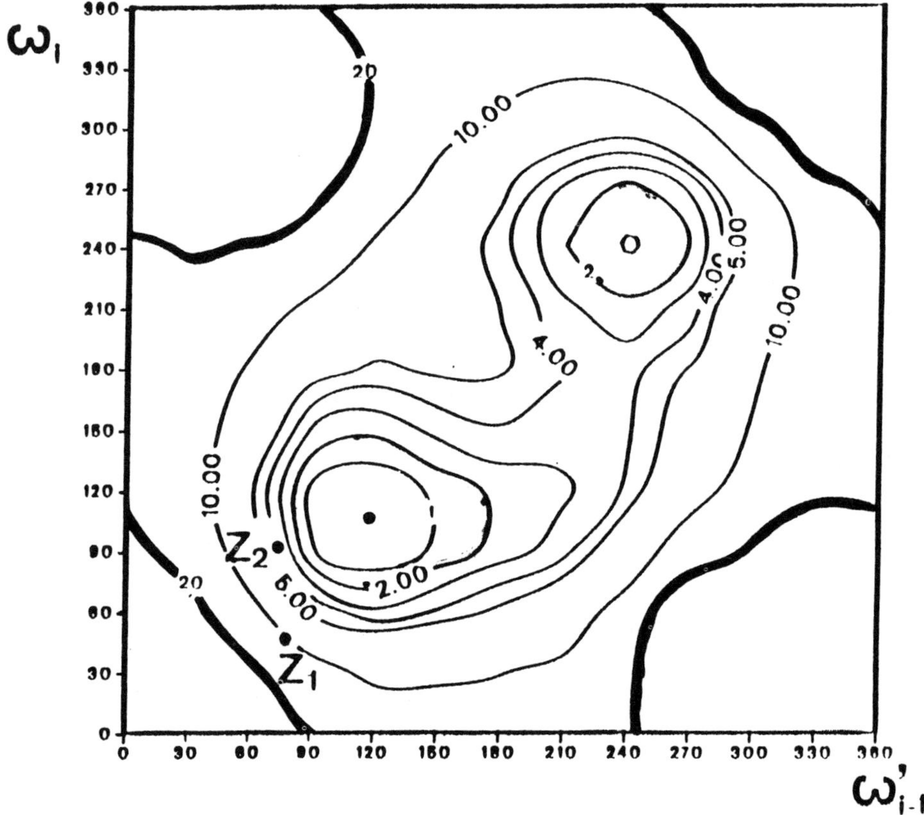

Fig 3: $E(\omega'_{i-1}, \omega_i)$ energetic map obtained with the $^2E^3E$ puckerings for the sugars and ψ_i in gt position. The 90°, 90° (g$^+$g$^+$) region is representative of the Z-DNA forms and the 270°, 270° (g$^-$g$^-$) one allows the access to the B forms. Comparison of this map with the equivalent obtained with ψ_i in gg position clearly shows that the transconformation $^2E^3E$gg --> $^2E^3E$gt induces the transconformation g$^-$g$^-$ --> g$^+$g$^+$ for the central phosphate group.

The Z_I et Z_{II} experimental stereostructures (respectively ω_{i-1}= 223° and 146°; ϕ_{i-1} = 221° and 164°; ψ_{i-1} = 56° and 66°; ψ'_{i-1} = 266° and 260°; ω'_{i-1} = 80° and 74°; ω_i = 47° and 92°; ϕ_i = 179° and 193°; ψ_i = 191° and 157°; ϕ'_i = 256° and 181°; ω'_i = 291° and 55°) are positioned on this map at about respectively 10 and 4 kcal/mol of the principal minimum.

On the basis of these results, we then conducted a minimization with Z_I and Z_{II} experimental structures as starting conformations for the computations. These computations lead to one, and only one, conformation (roughly corresponding, for the ω'_{i-1} and ω_i values, to the minimum minimorum of map of fig. 3) we call the Z-DNA optimized conformation illustrated in Fig. 4 and with the following rotational parameters values :

$$\phi_{i-1}= 160°;\ \psi_{i-1}= 64°;\ \phi'_{i-1} = 245°;\ \omega'_{i-1}= 120°;\ \omega_i= 102°;\ \phi_i = 171°;\ \psi_i = 178°;\ \phi'_i = 170°$$

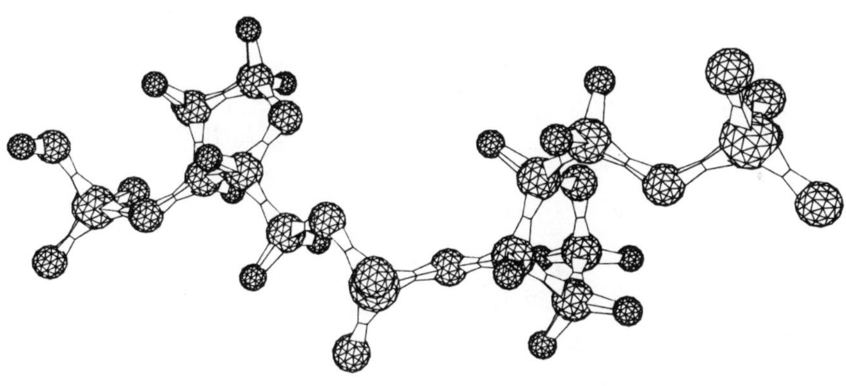

Fig.4 : perspective view of the Z-DNA optimized conformation obtained after minimization on all the rotational parameters.

So, on the basis of our theoretical conformational analysis, to the question "are Z_I and Z_{II} separated species?" it seems we can answer "no" and we can complete this information by the definition of the allowed range of conformations open to Z-DNA.

On another hand, to approach the possibilities of transconformations one of the major factors to examine is the rotation around the exocyclic bond (here, in our model ψ in the i unit) which induces a "kink" in the sugar-phosphate backbone.

In Fig. 5 we have reported the computed energy variation versus this ψ_i parameter (in the range $60°$ to $180°$) for various DNA forms for the two $(\omega'_{i-1}, \omega_i)$ regions about $(90°, 90°)$ and $(270°, 270°)$.

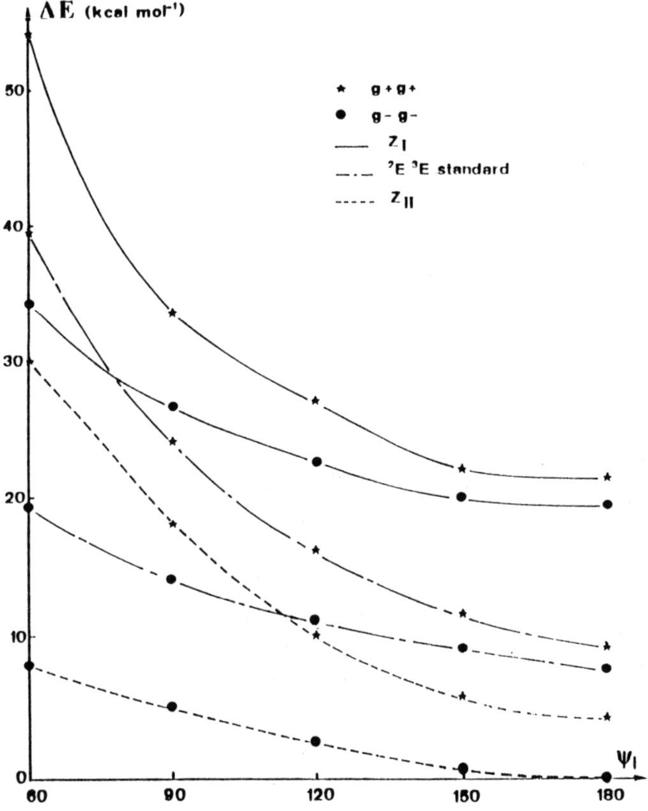

Fig.5: Energy variation versus ψ_i for various DNA forms for the two regions $90°, 90°$ and $270°, 270°$ of $(\omega'_{i-1}, \omega_i)$ respectively noted g^+g^+ and g^-g^-.

One of the major fact we can see on this figure is that the gap between the two $(\omega'_{i-1}, \omega_i)$ regions, which is in favor of the $(270°, 270°)$ one, decreases rapidly from about 20 kcal/mol to about 2 or 3 kcal/mol when changing ψ_i.

Clearly, the kink on ψ_i improves the accessibility of the $(90°, 90°)$ zone i.e. the global Z-DNA region and so, the possibility of a transconformation between a "standard" form and a Z one.

A lot of questions arise and one can ask about the validity of our approach using a model molecule "in vacuo" and not taking into account the water molecules around DNA. May be also Z structure will be instable in the absence of neutralizing cations (phosphate groups are very close in Z form).

In previous works on aminoacyladenylates, we have made computations on these molecules with added water (Broch and Vasilescu 1978, 1981). So, by the method we used to compute the energy of the molecule, we are already operational for the taking into account of hydration parts; but the problem is about the possibility of this procedure to take into account the *real* problem of water and not to hyper-simplify this one.

For example, the B --> Z transition can be induced in poly(d(A-T)).poly(d(A-T)) films by the presence of transition metal ions. Divalent transition metal ions as Co^{2+} or Ni^{2+} interact specifically with the N7 site of the purine so favoring the syn conformation of the base and, hence, inducing the Z structure (Adam et al. 1986). It seems that the combined effect of high sodium chloride and specific metal ions is necessary (Ridoux et al. 1988).
Concerning directly the problem posed, we can note that when the Z form of poly(d(A-T)).poly(d(A-T)), is observed a high structuration of the hydrating water molecules seems to be needed. The nickel ions, in water or in the solvent used by Liquier et al. (1984), are in the complex state $(Ni(H_2O)_6)^{2+}$. This hexaquo-nickel ion interacts with the N7 site of the adenine and may favor the syn geometry of the purine and thus the Z conformation of the polynucleotide.

So, if we want to treat correctly the problem, we have to take into account the real hydration scheme which is operating. This last imperative will be taken into account in future steps of our approach as we have done for the aa-AMP but the first *necessary* step is to determine the *intrinsic* allowed zones, the allowed stereostructures for the Z-DNA backbone solely, in the aim to minimize and render possible the exploration of the various parameters intervening in a following very time-consuming super-molecule (DNA plus "water" molecules) study.

REFERENCES

Adam, S., Liquier, J., Taboury, J.A., and Taillandier, E. (1986) Biochemistry **25**, 3220-3225
Arnott, S., Chandrasekaran, R., Millane, R.P., and Park H.S. (1986) Biophys. J. **49**, 3-5
Broch, H., and Vasilescu, D. (1978) Studia Biophysica, **1**, 21-28
Broch, H., and Vasilescu, D. (1981) in "Biomolecular Structure, Conformation and Evolution" (R. Srinivasan ed.), vol. **2**, Pergamon Press, pp. 329-343
Broch, H., and Vasilescu, D. (1989) Biopolymers, **28**, in press
Broch, H., Vasilescu, D., and Peronneau, H. (1977) Studia Biophysica **2**, 99-112
Broch, H., Viani, R., Grassi, H., and Vasilescu, D. (1989) Int. J. Quant. Chem., QBS **16**, in press

Crawford, J.L., Kolpak, F.J., Wang, A.H.J., Quigley, G.J., Van Boom, J.H., Van der Marel, G., and Rich, A. (1980) Proc. Natl. Acad. Sci. USA **77**, 4016-4020

Drew, H., Takano, T., Tanaka, S., Itakura, K., and Dickerson, R.E. (1980) Nature **286**, 567-573

Liquier, J., Bourtayne, P., Pizzorni, L., Sournies, F., Labarre, J.F., and Taillandier, E. (1984) Anticancer Res. **4**, 41-44

Pohl, F.M., and Jovin, T.M. (1972) J. Mol. Biol. **67**, 375-396

Ridoux, J.P., Liquier, J., and Taillandier, E. (1988) Biochemistry **27**, 3874-3878

Taboury, J.A., Liquier, J., and Taillandier, E. (1985) Can. J. Chem. **63**, 1904-1909

Wang, A.H.J., Quigley, G.J., Kolpak, F.J., Crawford, J.L., Van Boom, J.H., Van der Marel, G., and Rich, A. (1979) Nature **282**, 680-686

Wang, A.H.J., Quigley, G.J., Kolpak, F.J., Van der Marel, G., Van Boom, J.H., and Rich, A. (1981) Science **211**, 171-176

Water and Ions in Biomolecular Systems
Advances in Life Sciences
© 1990 Birkhäuser Verlag Basel

PROTEINS AS COUNTERIONS OF DNA: A NEW MODEL OF NUCLEOPROTAMINE STRUCTURE

Juan A. Subirana

Escola Técnica Superior d'Enginyers Industrials, Diagonal 647, 08028 Barcelona, Spain

SUMMARY: Both the C-terminal region of histone H1 and protamines have a tendency to acquire the α helical form in a large percentage. We suggest that such tendency is enhanced upon interaction with DNA. A new nucleoprotamine model based on this concept is developed, with the protamine molecules in a partial helical form running in parallel with the DNA molecules, each protamine binding together three DNA molecules. Similar structures may appear locally in the regions of interaction of DNA with histone H1 in chromatin.

INTRODUCTION

The phosphate charges of the DNA present in the nuclei of eukaryotic organisms are always neutralized by proteins. This neutralization is almost complete in those sperm nuclei which contain protamines, so that the basic amino acid/phosphate ratio is close to unity in such cells (Suau 1975; Ausió and Suau, 1983). This neutralization results in a complete insolubilization of the complex and only a relatively small change in its packing density is observed when the humidity of the sample is modified (Suau i Subirana, 1977). In fact a nucleoprotamine sample only accepts about 25% of its dry weight as water of hydration at 100% relative humidity.

In contrast, the DNA in somatic nuclei is not fully neutralized by histones. Each core histone has a net positive charge of about 20, so that the eight molecules in a nucleosome core only neutralize the equivalent of 80 base pairs of the 146 present in a core particle. In fact it has been shown that core particles can accept a much larger amount of histones (Voordouw and Eisenberg, 1978; Olivares et al, 1987). The addition of histone H1 does not allow complete neutralization of the complex either, as we will discuss in detail below. In fact external histones also bind to isolated chromosomes (Mirsky et al, 1968) or nuclei (Subirana, 1975). The relatively low degree of neutralization of DNA by histones makes the nucleohistone complex very sensible to the ionic conditions and to the presence of divalent ions.

In this paper we suggest that in spite of the charge differences, both protamines and part of histone H1 may interact in a similar way with DNA.

THE INTERACTION OF HISTONE H1 WITH DNA

It is generally accepted that the globular part of histone H1 interacts with the nucleosome core (Allan et al, 1980), whereas the highly charged C-terminal region neutralizes part of the phosphates found in internucleosomal DNA. We have recently analyzed the charge distribution in this histone region (Subirana, 1990). We have found that the percentage of positively charged residues varies significantly in different H1 histones, between 30 and 44%. In most cases these charges are uniformily distributed throughout the C-terminal region of histone H1. We conclude that this region may be considered as a featurless basic polypeptide with a rather constant charge density which reduces the charge of the nucleohistone complex in a smooth way, thus allowing the formation of a higher order structure in chromatin fibers.

Histone H1 has a variable net charge but always lower than 60^+. Assuming that there is one molecule per nucleosome, histone

H1 only neutralizes part of the charges of the DNA in the core and internucleosomal DNA, so that in the complete nucleohistone complex only about 60% of the phosphate charges in DNA are neutralized by histones. The other 40% are neutralized by inorganic and low molecular weight ions.

Of particular interest is the suggestion (Clark et al, 1988) that the C-terminal region contains helical regions, a question which we have discussed in detail elsewhere (Subirana, 1990). In fact we have found that some synthetic polypeptides rich in alanine and lysine do interact with DNA in a helical conformation (Azorín et al, 1985).

We suggest that the C-terminal region of histone H1 may interact with bundles of internucleosomal DNA and thus stabilize the higher order structure of nucleosomes in the 30 nm chromatin fiber, possibly in a zig-zag arrangement (Subirana et al, 1985). The main difference with the protamine molecules will be that the latter have a much higher charge density (60 to 80%) and allow a complete neutralization of DNA phosphates in the complex. Locally, the interaction may be quite similar in both cases, so that helical regions of histone H1 may interact with DNA in a way similar to that shown for protamine in Fig.1.

MODELS FOR NUCLEOPROTAMINE STRUCTURE

The DNA-protamine complex has a rather regular hexagonal packing arrangement for all the different protamines which have been investigated in spite of its differences in size, charge and amino acid composition (Suau and Subirana, 1977). However it is not clear whether protamines have any secondary structure in the complex and how they interact with DNA. In the first studies on nucleoprotamine structure (Feughelman et al, 1955; Suau and Subirana, 1977) it was assumed that protamine covers the narrow groove of DNA, on the basis of the strong intensity of the first layer line in the fibre diffraction pattern. This interpretation was challenged by Fita et al (1983), who compared the pattern given by DNA-arginine complexes with that of nucleoprotamines. They concluded that protamines are most likely located on the wide groove of DNA.

An atractive feature of either of these models is that they explain the hexagonal packing with screw disorder of DNA molecules, since the complex looks like a single helix rather than a double helix. Franklin and Klug (1956) had shown that single helices have a tendency to pack in a hexagonal way, whereas double helices show more complex packing arrangements (Marvin et al, 1961). Polyalanine, for example, in its helical form packs in a hexagonal cell (Brown and Trotter, 1956). On the other hand the above mentioned models imply that the protamine chain is almost fully extended (Fita et al, 1983; Puigjaner et al, 1986) and it is not clear how the peptide groups may form hydrogen bonds.

In protein structures most of the peptide groups form hydrogen bonds with other peptide groups. Therefore it is possible that in nucleoprotamines some sort of secondary structure develops in the protein so that intramolecular hydrogen bonds are formed. Actually Warrant and Kim (1978) suggested that protamine may adopt a bent helical form which follows one of the grooves of DNA. However this model does not appear feasible since the DNA in nucleoprotamines is tightly packed and there is not enough room to place an α helix on a DNA groove.

Secondary structure prediction methods applied to protamines (Toniolo, 1980; Cid and Arellano, 1982) conclude an abundance of β turns in these molecules. Such turns may in fact be present, but from their eventual presence it is not possible to deduce with certainty the overall shape of the protamine molecules when they interact with DNA.

AN ALTERNATIVE MODEL FOR NUCLEOPROTAMINE STRUCTURE

In this section we present a model for the nucleoprotamine complex in which the problems presented by the models we have

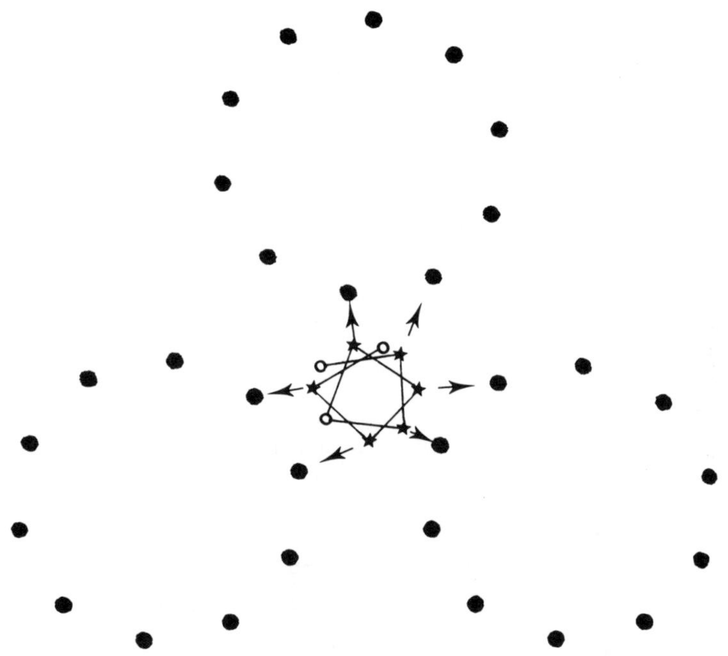

Fig.1. Projection of a model of nucleoprotamine. The black dots indicate the DNA phosphates placed at a radius of 9.5 Å. The DNA molecules have a center to center distance of 27.5 Å, similar to that found "in vivo" at high humidity. In the center a helical fragment of a protamine molecule, with nine amino acid residues is represented, with the sequence X-(Arg)$_4$ -X-(Arg)$_2$ -X, where X is an uncharged residue. This sequence is taken as an example, it is found in iridine (II) and in clupein Z. According to Toniolo (1980) it has the tendency to form a helical segment. In the figure the charged residues are indicated by stars and the uncharged residues are shown as empty circles. The symbols are placed on the position of the carbons of the amino acids, i.e., on the first atom of the side chains. The arrows indicate neighbouring phosphate charges which could be neutralized by the arginine residues in the helical segment shown in the figure. A similar arrangament might be present in some of the regions of interaction of the C-terminal part of histone H1 with internucleosomal DNA.

just discussed are not found. It is based on the assumption that a significant length of the protamine molecule (40-60%) may adopt a helical form upon interaction with DNA. Each protamine molecule would be placed among three DNA molecules, as shown in Fig.1, so that the axis of the protamine is roughly parallel to the axis of the DNA molecules. In turn, each DNA molecule would be surrounded by 6 protamine molecules. If we consider a hypothetical protamine with 30 amino acids and 20 charged residues, it should run about 2 turns of the DNA helix (68Å). Under these conditions a protamine molecule would neutralize 6.7 charges of each of the surrounding DNA molecules and vice versa, thus achieving complete charge neutralization. The protamine molecule would be about 50% helical and 50% partially extended, in order to span the required 68Å. Some β turns might be found in the regions between helical stems.

The following considerations favour this model:

a) Isolated protamines have a random coil conformation in solution, but upon the addition of chloroethanol they become partially helical (Bradbury et al, 1967; Toniolo et al, 1979). It is likely that upon neutralization of the arginine charges this tendency for a helical form will be enhanced. In fact secondary structure prediction favours the formation of α helical structures (Toniolo, 1980) in some regions of the protamine molecules.

b) Evolutionary considerations (Subirana and Colom, 1987) indicate a relationship between protamines and the C-terminal region of histone H1, while both have a tendency for the formation of helical regions (Clark et al, 1988; Subirana, 1990). The H1 histone found in the spermatozoa of some species has a very clear tendency for the α helical form (Subirana, 1990) and may be considered an intermediate step in the transition between somatic H1 and protamines.

c) The model would explain the high stability found in nucleoprotamine complexes, which are highly insoluble (Ausió and Suau, 1983) and difficult to dissociate.

d) The presence of the 001 reflection in fiber diffraction

patterns (Suau and Subirana, 1977) could be easily explained by the modulation every 34 Å of the protamine structure placed among three DNA double helices.

On the other hand it is not clear why the DNA maintains a certain degree of order in the vertical direction. It would appear that hexagonal packing should only be found in the equator, with random vertical displacement of the DNA molecules, as it is found in poly(Ala Lys) (Azorín et al, 1985). A more detailed analysis of the interactions may show why hexagonal packing with screw disorder is maintained in nucleoprotamines.

The model is consistent with the results found in complexes with synthetic polypeptides. The higher the charge of the polypeptide, the lower the amount of α helix which would be allowed by the model, since the polypeptide chain should become fully extended in order to neutralize the phosphate charges. In the case of DNA-polyarginine complexes (Suwalsky and Traub, 1972b) it is found that DNA is packed differently. It appears that under these conditions the polypeptide has a different conformation, perhaps wrapped around the DNA, but certainly different from that found in nucleoprotamines. In fact polyarginine under high humidity conditions looses its helical conformation to become fully extended (Suwalsky and Traub, 1972a).

At the other end, the regular polypeptide poly(Ala Lys), with a lower charge density (50%), acquires a high degree of helical conformation (Azorín et al, 1985). In this case the DNA charges can be neutralized by a 100% helical polypeptide. The observation of a very precise hexagonal order only in the equator favours the interpretation that the DNA molecules become then randomly displaced in the vertical direction (Azorín et al, 1985). It appears that the higher degree of order achieved in the complex with protamines may be due to local disruptions of the helical conformation, associated with the higher charge density usually found in these proteins.

Acknowledgment: This work has been suported in part by a grant from the Plan Nacional de Biotecnología BT87-0009.

REFERENCES

Allan, J., Hartman, P.G., Crane-Robinson, C., and Avilés, F.X. (1980) Nature 288, 675-679.

Ausió, J., and Suau, P. (1983) Biophys. Chem. 18, 257-267.

Azorín, F., Vives, J., Campos, J.L., Jordán, A., Lloveras, J., Puigjaner, L., Subirana, J.A., Mayer, R., and Brack, A. (1985) 185, 371-387.

Bradbury, E.M., Crane-Robinson, C., Goldman, H., Rattle, H.W.E., and Stephens, R.M. (1967) J.Mol.Biol. 29, 507-523.

Brown, L., and Trotter, I.F. (1956) Trans. Faraday Soc. 52 537-548.

Cid, H., and Arellano, A. (1982) Int.J.Biol.Macromol. 4, 3-8.

Clark, D.J., Hill, C.S., Martin, S.R., and Thomas, J.O. (1988) EMBO J. 7, 69-75.

Feughelman, M., Landgridge, R., Seeds, W.E., Stokes, A.R., Wilson, H.R., Hooper, C.W., Wilkins, M.H.F., Barclay, L.K., and Hamilton, L.D. (1955) Nature 175, 834-838.

Fita, I., Campos, J.L., Puigjaner, L.C., and Subirana, J.A. (1983) J.Mol.Biol. 167, 157-177.

Franklin, R.E., and Klug, A. (1956) Biochim. Biophys. Acta 19, 403-416.

Marvin, D.A., Spencer, M., Wilkins, M.H.F., and Hamilton, L.D. (1961) J.Mol.Biol. 3, 547-565.

Mirsky, A.E., Burdick, C.J., Davidson, E.H., and Littau, V.C. (1968) Proc.Natl.Acad.Sci., U.S.A. 61, 592-597.

Olivares, C., Azorín, F., Subirana, J.A., and Cornudella, L. (1987) Biophys. Chem. 28, 51-57.

Puigjaner, L.C., Fita, I., Arnott, S., Chandrasekaran, R., and Subirana, J.A. (1986) 3, 1067-1078.

Suau, P., (1975) Ph.D.Thesis, Faculty of Biology, University of Barcelona.

Suau, P., and Subirana, J.A. (1977) J.Mol.Biol. 117, 909-926.

Subirana, J.A. (1975) In: The Structure and Function of Chromatin (Fitzsimons, D.W. and Wolstenholme, G.E.W., Eds.) Elsevier/ North-Holland, Amsterdam, p. 153.

Subirana, J.A. (1990) Biopolymers, in the press.

Subirana, J.A., Muñoz-Guerra, S., Aymamí, J., Radermacher, M., Frank, J. (1985) Chromosoma 91, 377-390.

Subirana, J.A., and Colom, J. (1987), FEBS Letters 220, 193-196.

Suwalsky, M., and Traub, W. (1972a) Biopolymers 11, 623-632.

Suwalsky, M., and Traub, W. (1972b) Biopolymers 11, 2223-2231.

Toniolo, C., Bonora, G.M., Marchiori, F., Borin, G., and Filippi, B. (1979) Biochim. Biophys. Acta 576, 429-439.

Toniolo, C. (1980) Biochim. Biophys. Acta 624, 420-427.

Voordouw, G., and Eisenberg, H. (1978) Nature 273, 446-448.

Warrant, R.W., and Kim, S.H. (1978) Nature 271, 130-135.

EFFECT OF HYDRATION AND METAL IONS ON DNA CONFORMATIONS STUDIED BY VIBRATIONAL SPECTROSCOPY

E.Taillandier and J.Liquier

Laboratoire de Spectroscopie Biomoléculaire, UFR de Médecine, 74 rue Marcel Cachin F93012 Bobigny Cedex, France.

Water molecules and metal ions play a fundamental role in the stabilization of nucleic acid structures. It has been extensively studied both theoretically (for review see Pullman et al. 1983) and experimentally by X-ray diffraction (for review see Rich et al. 1984, Dickerson 1987) and NMR. We would like here to give a few examples of how vibrational spectroscopy can be used to investigate the effect of water and ionic environment on DNA conformation. Local changes of DNA geometry may play a fundamental role in regulation of gene expression for example by modulating DNA-protein interactions, therefore the importance of the studies concerning the conformational variability of the DNA double helix. One of the advantages of vibrational spectroscopy is to allow one to obtain informations about nucleic acid helical geometries under a wide variety of physical states: solutions, gels, fibers, hydrated films, crystals. It is therefore possible to study the effect of hydration on the geometry of a DNA sequence and investigate the existence of possible conformational transitions. Vibratioal spectroscopy is thus able to establish a relationship between the results obtained in solid state by X-ray crystal diffraction and the structures determined in solution by NMR studies.

B->A AND B->Z CONFORMATIONAL TRANSITIONS IN NUCLEIC ACIDS

Three main families of DNA structures are now classically considered: right-handed A and B, left-handed Z. Raman lines and infrared absorption bands characteristic of these nucleic acid conformations permit the identification of C3'-endo and C2'-endo families of nucleoside sugar puckers, of the _syn_ and _anti_ glycosil bond orientations and the differentiation between the A, B and Z backbone geometries. A normal coordinate analysis performed in parallel with experimental work allows us to assign the DNA marker modes and follow or predict the shifts of the corresponding spectral peaks in case of changes of the geometry of the molecule. Among the first infrared spectroscopy studies concerning the effect of dehydration on DNA conformation was the identification of the B->A transition in native DNAs induced by the decrease of the water content of nucleic acid hydrated films (Pilet & Brahms 1973). Since then this transition between two right-handed helical families induced in films by the decrease of relative humidity has been observed for sodium salts of synthetic polynucleotides, poly d(A-C).poly d(G-T) (Taillandier et al. 1984) poly d(A-T) (Taillandier et al. 1985) and oligonucleotides such as d(CCCGCGGG)$_2$ (Adam et al. 1989).

In the case of triple stranded poly dT.poly dA.poly dT such a decrease of hydration fails to induce a conformational transition. This triple stranded polymer can be easily characterized by infrared spectroscopy as shown in Figure 1. We present here the infrared absorption spectra recorded in D2O solutions of double stranded poly dA.poly dT (top) and triple stranded poly dT.poly dA.poly dT (bottom). The spectral region displayed corresponds to the absorptions of the in-plane double bond stretching vibrations of the bases. It is extremely sensitive to base pairing, and we can observe that the infrared band located at 1622 cm^{-1} in the poly dA.poly dT spectrum (top) and assigned mainly to a vibration of the N7=C8 double bond is no longer detected on the spectrum of poly dT.poly dA.poly dT (bottom), clearly showing the formation of a Hoogsteen type base pairing between the dA and the second dT strands. The effect of the decrease of the relative humidity on a poly dT.poly dA.poly dT film is shown in Figure 2. This spectral region (between 1000 and 750 cm^{-1}) presents mainly vibrations of the deoxyribose-phosphate chain. The heterogeneity of the deoxyribose conformation is clearly detected by the simultaneous presence of absorptions located around 867, 839 and 823 cm^{-1}. However very little significative modifications are detected when the relative humidity is decreased even to 32% R.H., showing the stability of this triple helix.

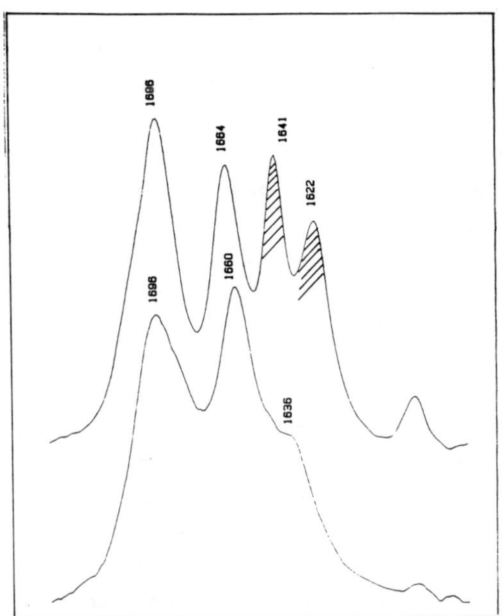

Figure 1. Infrared absorption spectra in D2O solutions of:
top: poly dA.poly dT; bottom: poly dT.poly dA.poly dT

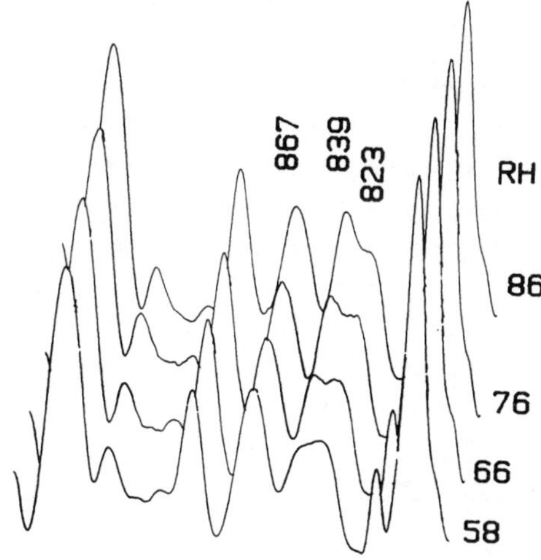

Figure 2. Infrared absorption spectra between 1000 and 750 cm^{-1}
of a poly dT.poly dA.poly dT film at decreasing relative
humidities

The decrease of the relative humidity in the case of an alternating poly d(G-C) sequence induces a completely different conformational transition, from a right-handed B helix to a left-handed Z helix (Taillandier et al.1981, Taboury et al. 1985).

The induction of this B->Z transition is more difficult to achieve when A-T base pairs are incorporated in alternating G-C sequences. As previously stated the decrease of the relative humidity of a poly d(A-C).poly d(G-T) film film will result in a B->A transition. The simple decrease of water activity obtained either by dehydration in films or by increase of ionic strength in solution fails to induce the Z form of the polymer. However if nickel ions are added in these low water activity conditions, poly d(A-C).poly d(G-T) does adopt a Z family conformation (Taillandier et al. 1984). It has been shown that the nickel hydrated ions interact preferentially with the purine bases on the N7 site of guanines and adenines, stabilizing the _syn_ conformation of the purine nucleosides and therefore the Z geometry of the DNA (Adam et al. 1986a).

The decrease of water activity combined with specific interactions with nickel ions can be used to induce the B->Z transition in a polynucleotide containing only A-T base pairs, poly d(A-T). The left-handed Z form of poly d(A-T) has been characterized by infrared absorption (Adam et al. 1986b), UV circular dichroism (Bourtayre et al. 1987) and Raman spectroscopy (Ridoux et al. 1988).

This effect of nickel ions has also been observed in the case of oligonucleotides such as d(CCCGCGGG)$_2$ which undergoes a B->A transition by decrease of the relative humidity in presence of sodium ions, but adopts a Z form in presence of nickel ions (Adam et al. 1989).

The characterization of the left-handed helixes of poly d(A-C).poly d(G-T) and poly d(A-T) gives us a set of infrared marker bands and of Raman marker lines for G-C as well as A-T base pairs and for the phosphodiester chain which are remarquably similar for all regularly alternating purine-pyrimidine sequences, which leads us to propose that the Z geometry has certainly standard non base specific characteristics.

<u>VIBRATIONAL SPECTROSCOPY OF CRYSTALLIZED OLIGONUCLEOTIDES</u>

Instead of variing the water activity by changing the relative humidity of a film it is possible to study the effect of hydration by comparing the conformations adopted by a DNA sequence in solution and in crystal.

Recent progress in spectroscopic techniques allows us now to obtain Raman and FTIR spectra of crystals using microscope attachments. The crystallization often induces a change in the geometry of the oligonucleotide and this is the case for example for the d(CACGTG)$_2$ sequence. The microRaman spectra clearly show that this oligonucleotide adopts a B geometry in solution and a Z geometry in the crystal (Urpi et al. 1989). However we do not always observe a conformational transition occuring under the cristallization process. For example the microRaman spectrum of the d(GGCGCC)$_2$ crystal reflects a perfectly classical B type conformation (Figure 3 top). The reverse 5′-3′ sequence d(CCGCGG)$_2$ has a more peculiar behaviour. The FTIR spectra of d(CCGCGG)$_2$ hydrated films show the existence of a B->Z transition obtained by decrease of the water content. Similarly this transition is observed in solution by increasing the ionic strength (Figure 4 top and bottom). One would therefore expect that the d(CCGCGG)$_2$ sequence should adopt a Z geometry in the crystalline state. However the microRaman (Figure 3 bottom) and microFTIR (Figure 4 center) spectra clealy show that there is no Z geometry in the crystal. The conformation is mainly right handed with slight modifications when compared to a cannonical B form.

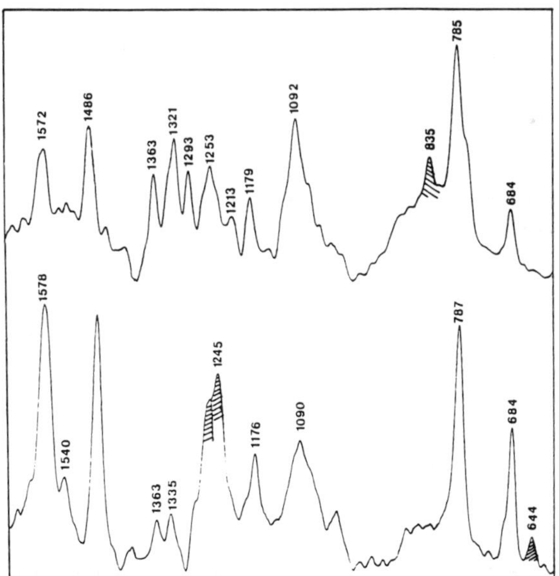

Figure 3. MicroRaman spectra of oligonucleotide crystals.
Top: d(GGCGCC)$_2$. Bottom: d(CCGCGG)$_2$.

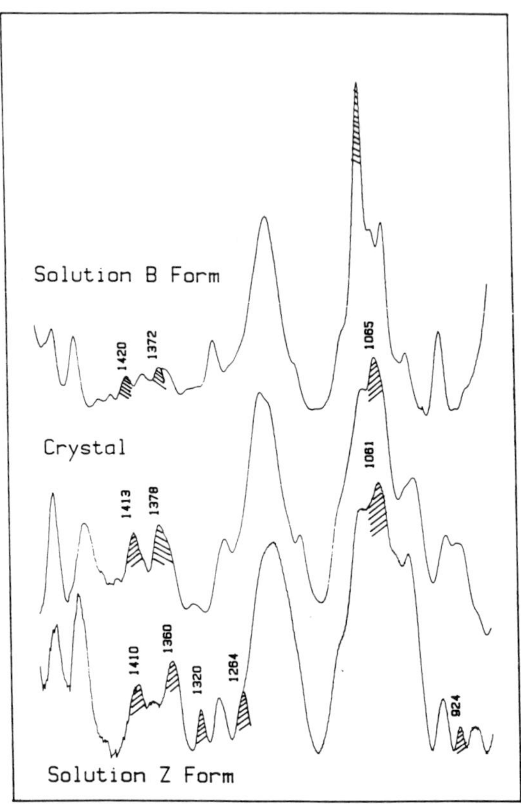

Figure 4. FTIR spectra of d(CCGCGG). Top: low salt solution.
B form. Center: crystal, intermediate form. Bottom: high salt
solution, Z form.

Mixtures of DNA geometries can be detected using the marker peaks of the different conformations. An example of such application is given below. The FTIR spectrum of the d(GGTATACC)$_2$ crystal presents in the region between 1000 and 750 cm^{-1} and involving mainly the vibrations of the deoxyribose-phosphate chain two bands located at 861 and 811 cm^{-1} (Figure 5 bottom) characteristic of an A type geometry, which indeed is the structure found by X-ray diffraction studies (Shakked et al. 1983). However besides these expected A form bands we observe a weaker band at 834 cm^{-1} which indicates the presence of some B type geometry in the crystal. This is in excellent agreement with the recently published structure of a closely related sequence d(GGBrUABrUACC)$_2$ in which both A and B geometries were shown to exist in the crystal lattice (Doucet et al.1989).

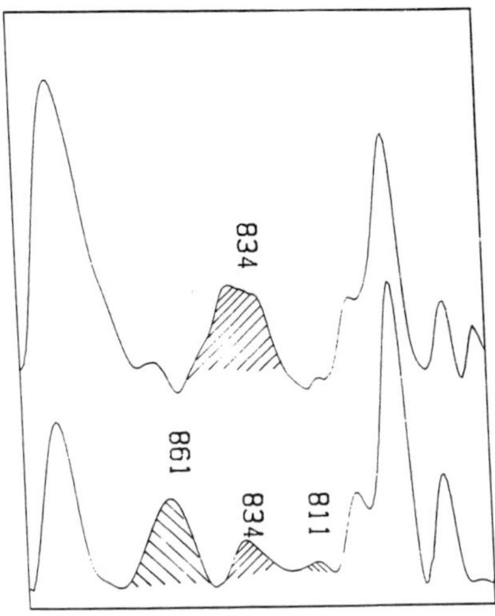

Figure 5. MicroFTIR spectra of d(GGTATACC). Top: solution
Bottom: crystal. Marker bands of: B form   ,A form

REFERENCES

Adam, S., Bourtayre, P., Liquier, J. and Taillandier, E. (1986a)
   Nucleic Acids Res. 14, 3501-3513.
Adam, S., Liquier, J., Taboury, J.A. and Taillandier, E. (1986b)
   Biochemistry 25, 3220-3225.
Adam, S., Peticolas, W., Huynh-Dinh, T., Igolen, J. and Taillandier, E.
   (1989) Spectrochim. Acta in the press.
Bourtayre, P., Liquier, J., Pizzorni, L. and Taillandier, E. (1987)
   J. Biomol. Struct. Dyn. 5, 97-104.
Dickerson, R.E. in Unusual DNA Structures (R.D. Wells and S.C. Harvey Eds.)
   Springer Verlag.
Doucet, J., Benoit, J.P., Cruise, W.B.T., Prangé, T. and Kennard, O.
   (1989) Nature 337,190-193.
Pilet, J. and Brahms J. (1973) Biopolymers 12, 387-403.
Pullman, B., Pullman, A. and Lavery, R. in Structure Dynamics Interactions
   and Evolution of Biological Macromolecules (C. Helene Ed.)
   Reidel Pub. pp 23-44.
Rich, A., Nordheim, A. and Wang, A.H-J (1984) Ann. Rev. Biochem. 53,
791-846.
Ridoux, J.P., Liquier, J. and Taillandier, E. (1988)
   Biochemistry 27, 3874-3878.

Shakked, Z., Rabinovitch, D., Kennard, O., Cruise, W.B.T., Salisbury, S.A.
   and Viswamitra, M.A. (1983) J. Mol. Biol. 166, 183-201.
Taboury, J.A., Liquier, J. and Taillandier, E. (1985)
   Can. J. Chem. 63, 1904-1909.
Taillandier, E., Taboury, J.A., Liquier, J., Sautiere, P.and Couppez, M.
   (1981) Biochimie 63, 895-898.
Taillandier, E., Taboury, J.A., Adam, S. and Liquier, J. (1984)
   Biochemistry 23, 5703-5706.
Taillandier, E., Liquier, J. and Taboury,J.A. (1985) in Advances in
   Infrared and Raman Spectrosopy (R.J.H. Clark and R.E. Hester Eds.)
   Wiley, vol 12,pp 65-114.
Urpi, L., Ridoux, J.P., Liquier, J., Verdaguer, N., Fita, I.,
   Subirana, J.A., Iglesias, F., Huynh-Dinh, T., Igolen, J.
   and Taillandier, E. (1989) Nucleic Acids Res. 17, 6669-6680.

SHOWING - UP THE Na+ ENTROPY OF FLUCTUATIONS DURING THE DNA THERMAL TRANSCONFORMATION BY ^{23}Na NMR : INFLUENCE OF RADIOPROTECTOR CYSTEAMINE.

D. VASILESCU, J. LEMATRE and G. MALLET

Laboratoire de Biophysique, Université de Nice-Sophia Antipolis, Parc Valrose 06034 Nice Cedex, France.

SUMMARY. The " entropy of fluctuations S_f " concept, developed by Lenk, is applied to Na+ counterions of DNA during thermal transconformation. By using the ^{23}Na NMR technique, DNA thermal transconformation is studied in presence or absence of cysteamine radioprotector.

THE ENTROPY OF FLUCTUATIONS CONCEPT.

In biophysical systems, the molecular movements are important, so their internal energy is a function of fast molecular kinetic energy. Lenk (Lenk, 1979), assuming that entropy is mainly related to these movements, introduced the concept of " entropy of fluctuations S_f " (dynamic entropy) :

" *Unoriented or partially oriented samples (polymers, liquid crystals, biological systems), characterized by the significant molecular movements, can be sometimes described by the statistical parameters only. Furthermore, internal energy in this case is given by the kinetics of the molecular motion and the most significant contribution to entropy is given by the dynamic component, called " entropy of fluctuations ", which is an explicit function of the correlation time,* τ_c (tLenk, 1986).

Lenk demonstrated that S_f is inversely proportional to the temperature - dependent correlation time τ_c of Brownian microfluctuations :

$$S_f \, \alpha \, \tau_c^{-1} \quad (1)$$

In the case of transitions between two different states A and B, relation 1 gives :

$$Sf_A \ / \ Sf_B \ = \ \tau_c{}^B \ / \ \tau_c{}^A \qquad (2)$$

Then, according to James (James, 1975), we can introduce the spin-lattice relaxation time T_1 as a function of the spectral density $J(\omega)$ by the following relation :

$$1 \ / T_1 \ = \ K < H^2 > J(\omega) \quad (3)$$

where : K is a constant, $< H^2 >$ the mean square value of the local magnetic field and ω the rotational pulsation. Generally, $J(\omega)$ is correlated with "motional" behaviour, and $< H^2 >$ to "structural" behaviour. So, as a result of the "small-step" diffusion models, T_1 can be related to the correlation time τ_c by the equation:

$$1 \ / \ T_1 \ \alpha \ J(\omega) \ \approx \ \omega^{-m} \ \tau_c{}^{(1-m)} \qquad (4)$$

where m is a parameter that takes the values 0 or 2 in the case of rotational spectral density, and m = 1/2 or 3/2 when one deals with translational spectral densities.
Substituting (4) into (2), we obtain:

$$Sf_A \ / \ Sf_B \ = (T_1{}^A / \ T_1{}^B)^{1/(1-m)} \qquad (5)$$

The aim of this study is to investigate the variation of Na^+ entropy of fluctuations during the DNA thermal transconformation by using the ^{23}Na NMR technique in presence or absence of the radioprotector cysteamine.

MATERIALS AND METHODS.

DNA solutions were prepared from sodium salt of highly polymerized DNA obtained from calf thymus (Sigma, type D1). The DNA was slowly dissolved in NaCl $10^{-2}M$, prepared from Merk "suprapur" NaCl. The measured pH was 6.0. Sodium contents of solutions were determined by using atomic emission spectroscopy (Pye Unicam SP9). DNA concentrations expressed in phosphate sites (P) and melting profiles were determined with a Gilford spectrophotometer 2600 at 258 nm. An extinction coefficient of 6700 ± 100 mol^{-1} $l.cm^{-1}$ was used. Hyperchromic factors of DNA solutions were always greater than 30%. The lack of proteins was checked by the measurement of the optical absorbance ratio $A_{280} / \ A_{260}$ which was always lesser than 0.60 indicating that the material was essentially free of proteins.

The ^{23}Na NMR experiments were performed at 23.81 MHz on a Brücker WH-90/DS spectrometer operating in the pulsed mode with a single coil configuration. The 10 mm tubes each containing 2.0 ml of DNA solution, were put inside a 15-mm tube containing deuterium ^{2}H$_2$O, the resonance of which was used for the lock signal. The field homogeneity was optimized by means of magnet shim controls before each series of accumulations. The shim settings were adjusted until the linewidth of ^{23}Na in a standard sample of 10^{-1} M NaCl was no greater than 7.5 Hz. Each measurement was made with 10 μs long pulses and 5 μs delays. Eleven thounsand free induction decays (FID's) were accumulated in the 16k channels of the memory with a 3012 Hz sweep width. Sample temperature was carefully monitored within 0.2 °C range with a Brücker BST 100 / 700 temperature controller.

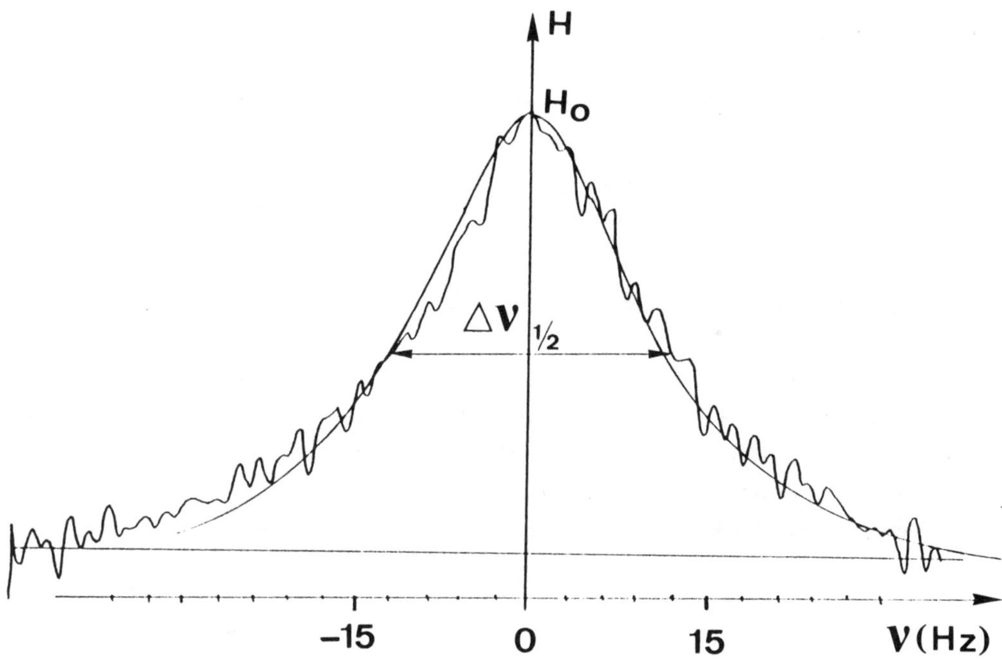

Figure 1. The experimental ^{23}Na NMR peaks exhibit Lorentzian shape.
(full line: the experimental curve; dotted line: the theoretical line constructed from the maximum amplitude H$_0$ and the linewidth $\Delta v_{1/2}$ of the experimental curve).

DNA in solution must be considered like a dynamic molecule, the structure and the behaviour of which depending on external parameters such as temperature, solvent, Na^+ counterions and magnetic field. When we work, as we did, in the "narrowed region" ($\omega_o^2 \tau_c^2 \ll 1$, in which ω_o is the Larmor pulsation) the ^{23}Na line shape is Lorentzian as shows figure 1. It follows that the linewidth $\Delta\nu_{1/2}$ of the peak at half height is a direct measurement of the ^{23}Na ions transverse relaxation rate R :

$$R = \pi\Delta\nu_{1/2} = \frac{1}{T_2} = \frac{1}{T_1} \quad (6)$$

since in this case, T_1 is equal to the spin - spin relaxation time T_2 .

We shall interpret the previous relations in the two following cases: m = 0, as it is generally admitted and, m = 1/2 as found by Lenk (Lenk, 1986) in the case of certain biophysical phase transitions. So, relations (3) and (5) become respectively :

$$1 / T_1 \, \alpha \, \tau_c \quad \text{and} \quad Sf_A / Sf_B = (\Delta\nu_{1/2}{}^B / \Delta\nu_{1/2}{}^A) \qquad (7) \text{ for m = 0}$$

$$1 / T_1 \, \alpha \, \tau_c{}^{1/2} \quad \text{and} \quad Sf_A / Sf_B = (\Delta\nu_{1/2}{}^B / \Delta\nu_{1/2}{}^A)^2 \qquad (8) \text{ for m = 1/2}$$

DNA THERMAL TRANSCONFORMATION.

As proposed by Bleam et al (Bleam et al., 1980), it is reasonable to admit that sodium ions of the solution are divided into two magnetically different classes: " bound " ions whose distance from the phosphate sites is within a few hydrated ionic radii, and " free " ions which are farther away and are not affected by field gradients arising from the polyanion charged groups. In our experimental conditions, the " fast - exchange " condition is fulfilled, so we can write :

$$R = p_B R_B + p_F R_F \quad (9)$$

where p_B and p_F are the fractions of bound and free sodium ions, R_B and R_F are the bound and free sodium relaxation rates, with:

$$p_B + p_F = 1 \qquad (10)$$

We propounded (Lematre et al., 1989) to interpret DNA thermal transconformation by introducing an appropriate model for two conformational states (helix (h) and coil (c)):

$$R^h = p_B^h R_B^h + p_F^h R_F^h \qquad (11)$$

$$R^c = p_B^c R_B^c + p_F^c R_F^c \qquad (12)$$

where R^h and R^c are the measured rates in the helix and coil states respectively.

Figure 2 shows the average results of several experiments: curve 1 is the variation of ^{23}Na linewidth versus temperature and curve 5 the behaviour of 10^{-2} M NaCl solution in which all Na ions are free. So, it is clear that when DNA thermal transconformation is achieved, all the Na ions are free, since curve 1 joins curve 5. This convergence of R^c towards R_{NaCl} may be interpreted in the following manners :

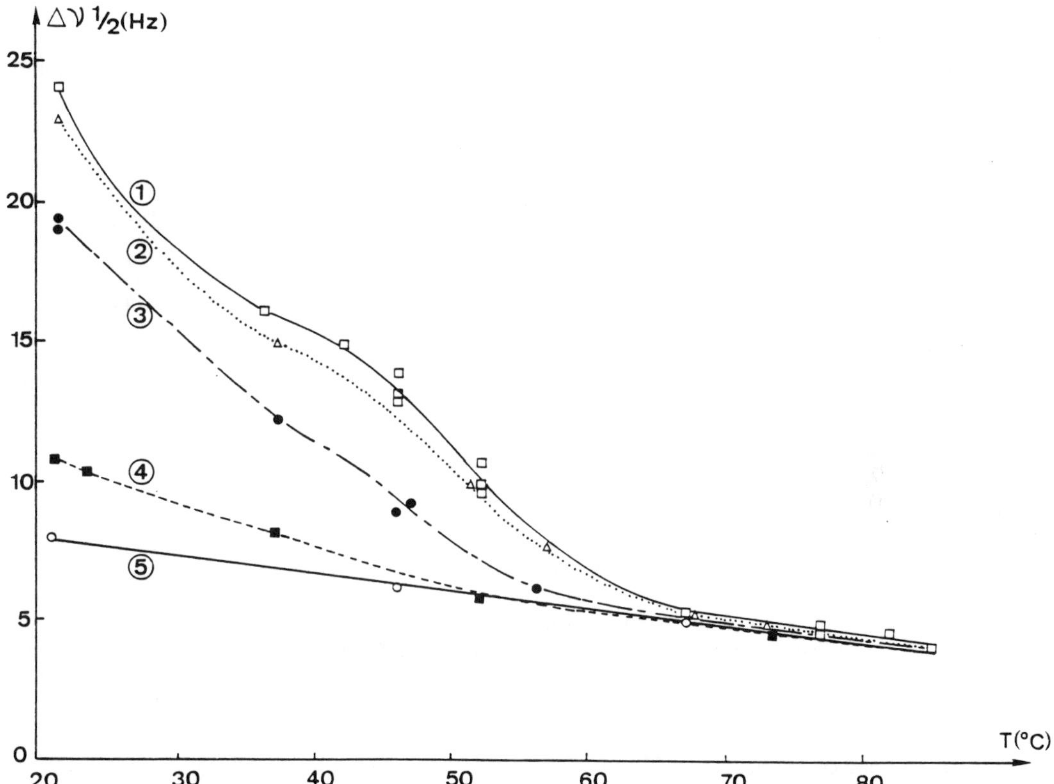

Figure 2. ^{23}Na linewidth measurements of Na-DNA versus temperature for different concentrations of added cysteamine. (drug concentrations are expressed in molar phosphate sites P of DNA).
1. Native DNA (10^{-2}M) in NaCl 10^{-2}M.   2. DNA plus cysteamine P/10.   3. DNA plus cysteamine P/4. 4. DNA plus cysteamine P. 5. Solvent NaCl 10^{-2}M.

1°. The product $p_B^c R_B^c$ of relation (12) tends to 0 with $p_B^c \ddagger 0$ since it is not temperature - dependent (Bleam et al. 1983); thus, R_B^c must also tend to 0. We deduced from our experiments that p_B^c is of the order of 0.1.

2°. The field of the coil being relatively weak, it is reasonable to admit that : $R_B^c = R_F^c$. It can therefore be deduced from relations (10) and (12) that :

$$R^c = R_B^c = R_F^c$$

This affirmation is corroborated by the behaviour of curves 1 in figures 3.

The narrowing of the ^{23}Na linewidth during the DNA thermal transconformation proves that the resulting ejection of sodium ions reduces the polyion mean charge density and thus diminishes its electric field. There is consequently a reduction of the screened coulombic potential and, as previously reported (Braulin et al., 1986), a delayed radial diffusion of sodium ions near the phosphate sites, inducing a decrease of the correlation time which is proportional to $1 / T_2$.

In an attempt to interpret the DNA thermal transconformation by using the concept of entropy of fluctuations, we reported figures 3a (m = 0) and 3b (m = 1/2), the normalized Na^+ entropy of fluctuations ratios of DNA with regard to NaCl solvent, in function of temperature. This representation illustrates the behaviour of "Na^+ ion entropy of fluctuations" during DNA thermal transconformation.

Curve (1) relative to DNA alone shows - independently of the m value - :

- a sigmoidal behaviour similar to that observed during spectrophotometric measurements ;

- that as long $S_{DNA} / S_{NaCl} < 1$, Na^+ ions are better ordered in DNA solution than Na^+ ions in NaCl solvent where they are free ;

- that in the coil state, all the Na ions are " free " with regard to the ^{23}Na NMR experiment. This assertion is in agreement with the previous relation : $R_B^c = R_F^c$.

When S_{DNA} / S_{NaCl} tends to 1, the DNA thermal transconformation is achieved, and Na^+ ions are in the same disorder in the DNA solution as in the NaCl solvent : in this case Na^+ ions behave as if they were free.

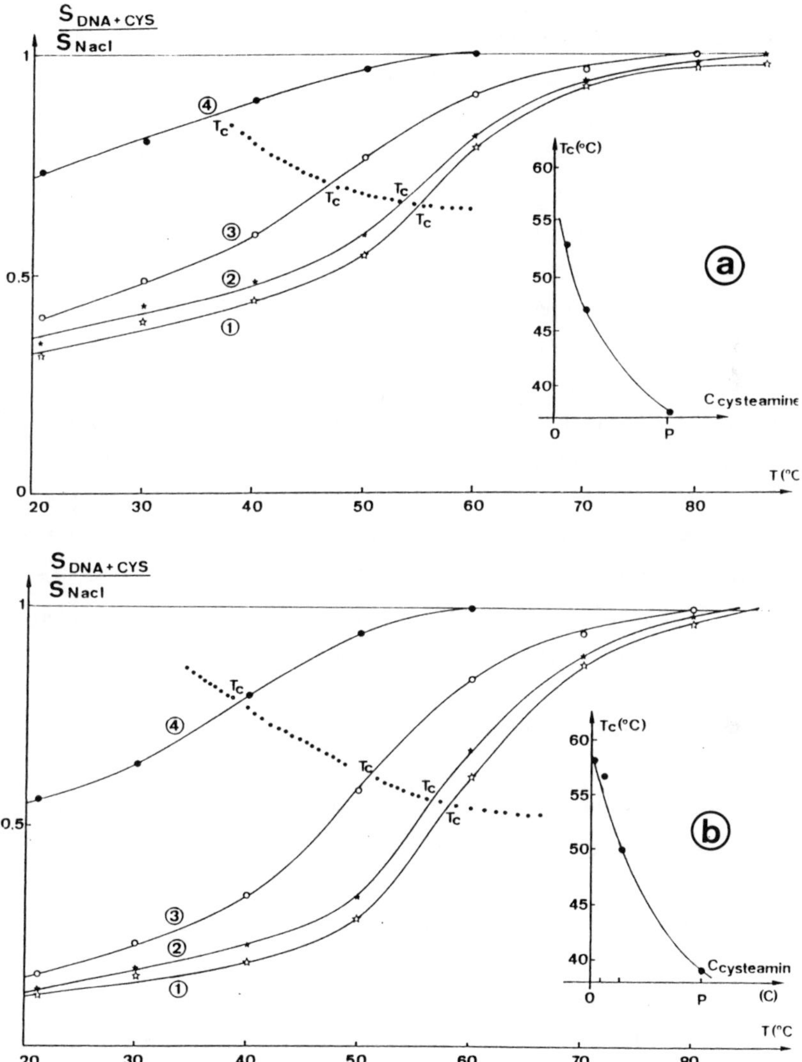

Figures 3a (m = 0) and 3b (m = 1/2) : Normalized Na$^+$ entropy of fluctuations ratios of DNA - cysteamine mixtures versus NaCl in function of temperature.
1. Na$^+$ entropy of fluctuations in DNA alone. 2. Na$^+$ entropy of fluctuations in DNA + cysteamine (P/10). 3. Na$^+$ entropy of fluctuations in DNA + cysteamine (P/4). 4. Na$^+$ entropy of fluctuations in DNA + cysteamine (P), (S$_{NaCl}$) is the Na$^+$ entropy of fluctuations in NaCl at the same temperature T).

THE RADIOPROTECTOR EFFECT.

INTERACTION OF DNA WITH CYSTEAMINE AT A FIXED TEMPERATURE OF 25°C,

It is well known that aminothiol compounds are radioprotector agents (Foye, 1973; Bacq, 1975; Yuhas,1980; Nygaard and Simic, 1983). Cysteamine - which is the decarboxylated cysteine - characterized by a reduction factor DRF = 1.6 , is a good radioprotector " in vitro " but introduces an unacceptable toxicity. Purdie (Purdie, 1971) has demonstrated that, in vitro, cysteamine penetrates the human cell and interacts with the nucleus. Thus, it appears that DNA is the most important target for this molecule. At the molecular level, the repair mechanism of DNA single and double strands broken by ionizing radiations involves the aminothiol as radical scavengers and as hydrogen donators which are able to repair the sugar damage (C4' radical in DNA deoxyribose) (Ward 1975, 1983; Schulte - Frohlinde, 1983). In absence of ionizing radiation, the interaction between the metabolic acting drug and the accessible portions of DNA is an important factor. We demonstrated previously that, in vitro, this interaction is essentially electrostatic and involves the cationic groups of the aminothiols and the anionic DNA phosphate sites (Mallet , 1982 - 1987). We showed that, by using " ab initio " quantum molecular computations (Broch et al., 1980 - 1982), the charge of the monocation (NH_3^+- (CH_2)$_2$ - SH) is distributed over the ammonium group (+ 0.516), the methylenic group adjacent to NH_3^+ (+ 0.299) and over the sulfhydryl group (+ 0.194). This interaction is accompanied by the ejection of a fraction of " bound " Na^+ out of the cylindrical sheath containing the condensed counterions.

We demonstrated this effect (Vasilescu and Mallet, 1985) by a ^{23}Na experiment at a fixed temperature of 25°C. In this case (fig. 4), the linewidth $\Delta v_{1/2}$ of ^{23}Na spectra diminishes dramatically as the cysteamine concentration is increased . For cysteamine concentrations lower than 3P/4, the linewidth drops from 23.5 to 11.5 Hz, showing a strong interaction of cysteamine with DNA.

Beyond this concentration, the linewidth remains constant, there being a saturation effect. This behaviour has been interpreted using a modified two states model and Manning's theory (Manning, 1969 - 1972- 1978) under rapid exchange conditions. We concluded that the addition of cysteamine to DNA solutions has conformational effects that can be attributed to a molecular rigidification resulting from bridging of two DNA phosphate sites by the radioprotective drug.

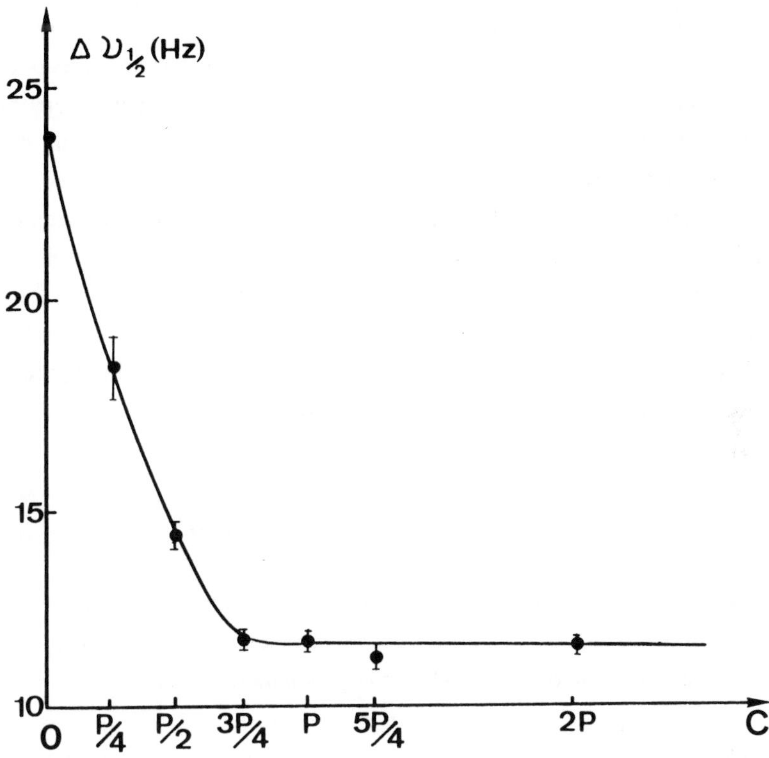

Figure 4. ^{23}Na Linewidth measurements of Na-DNA (10-2M) as a function of added cysteamine (concentration expressed in molar phosphate sites P of DNA.

Na$^+$ ENTROPY OF FLUCTUATIONS BEHAVIOUR DURING THERMAL TRANSCONFORMATION

For DNA radioprotected by cysteamine , the figure 3 shows (curves 2, 3 and 4) that:

$$\frac{S_{DNA + Cyst}}{S_{NaCl}} > \frac{S_{DNA}}{S_{NaCl}} \qquad (13)$$

88

In this case, for Na^+ ions, disorder appears to increase because the addition of radioprotector enhances the release of Na^+ ions from DNA phosphate sites. The effect of cysteamine is very sensitive at 21°C, where we observe a shift in the curves, which is - as expected - a function of the drug concentration. Consequently, the amplitude variation between the two levels of the sigmoids is less marked when DNA is radioprotected than when it is native. The same figure also gives the variations of the "critical temperature T_c" corresponding to the half of the amplitude variation versus cysteamine concentration. This uniform decrease of T_c, as cysteamine concentration increases, is characteristic of the fact that Na^+ ions ejection is enhanced by cysteamine addition . The graph utilized appears to amplify this effect too. On the other hand, at 21°C, there is a greater apparent order for Na^+ ions in DNA solutions for $m = 1/2$ than for $m = 0$.

In order to obtain a convenient graphic representation correlating the radioprotector effect at the macromolecular level with Na^+ entropy of fluctuations variations in DNA-cysteamine complexes as temperature T varies, we introduced (figure 5) the parameter x :

$$x = \frac{(S/S_{NaCl})T}{(S/S_{NaCl})21°C} - 1 \qquad (14)$$

where S is the Na^+ entropy of fluctuations in the DNA-cysteamine complex at temperature T and S_{NaCl} is the Na^+ entropy of fluctuations in NaCl at the same temperature, $(S/S_{NaCl})21°C$ being the Na^+ entropy of fluctuations ratio of the complex to NaCl solution at 21°C.

In addition to the evident decrease of the magnitude of x as the cysteamine concentration increases, which is characteristic of the order induced by the cysteamine radioprotector, we may notice :

- the presence of an isosbestic point I at 56°C ;

- that above the isosbestic point I, the x amplitude for native DNA is always superior at the one for DNA-cysteamine complexes, proving that above 56°C, the ejection of ^{23}Na is so important that the interaction of cysteamine drug with phosphate sites is enhanced, providing an increase in the rigidification of the DNA molecule.

In conclusion, with the help of ^{23}Na NMR technique, we have demonstrated for the first time that the "entropy of fluctuations" concept, applied to Na^+ counterions of DNA solutions, is suitable for understanding the thermodynamical behaviour of DNA thermal transconformation. It is also to be noted that, in our measurements, the values of the m parameter do not seem to affect the

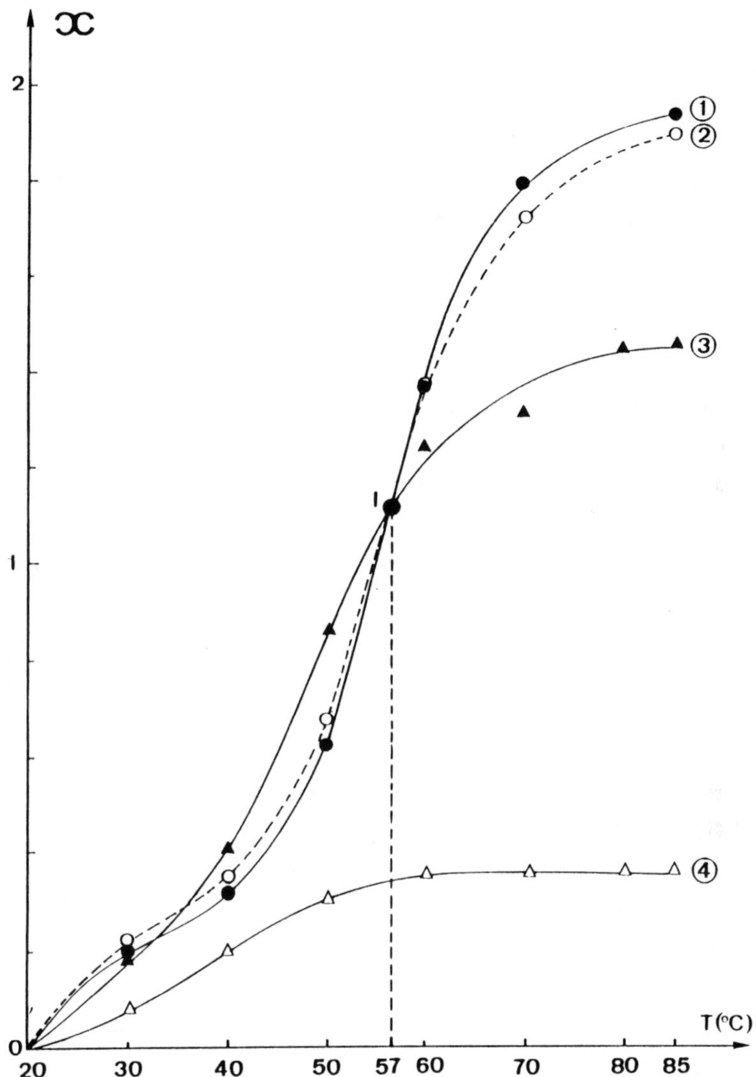

<u>Figure 5 (m = 1/2)</u>. Na$^+$ entropy of fluctuations variations in DNA-cysteamine mixtures versus temperature. $x = \dfrac{(S/S_{NaCl})_T - (S/S_{NaCl})_{21°C}}{(S/S_{NaCl})_{21°C}}$.

1. S = Na$^+$ entropy of fluctuations in DNA alone. 2.S = Na$^+$ entropy of fluctuations in DNA + cysteamine (P/10). 3. S = Na$^+$ entropy of fluctuations in DNA + cysteamine (P/4). 4. S = Na$^+$ entropy of fluctuations in DNA + cysteamine (P). (S$_{NaCl}$ is the Na$^+$ entropy of fluctuations in NaCl at the same temperature T, (S$_{NaCl}$)$_{21°C}$ being the Na$^+$ entropy ratio of the complex by NaCl solution at 21°C).

curves significantly; nevertheless, it appears that a rotational description is better suited to an apparent disorder than a translational description. Moreover, the radioprotector effect of cysteamine was well apprehended at the molecular level.

Aknowledgments: This research was supported by Grant DRET (Project 89/089). The authors wish to thank Professor G.S. Manning (Rutgers University) for stimularing discussions.

References

Bleam, M.L., Anderson C.F. and Record M.T.J., (1980) Proc. Natl. Acad. U.S.A. 77 3085 - 3089.

Bleam, M.L., Anderson C.F.and Record M.T.J., (1983) Biochemistry 22 5418 - 5425.

Braulin, W.H., Anderson C.F.and Record M.T.J. , (1986) Biopolymers 25 205 - 214.

Broch, H., Cabrol D. and Vasilescu D., (1980) Int. J. Quant. Chem., Quant. Biol. Symp. 7 283 - 295.

Broch, H., Cabrol D. and Vasilescu D. , (1982) Int. J. Quant. Chem., Quant. Biol. Symp.9 111 - 123.

James, T.L., (1975) Nuclear Magnetism in Biochemistry, Acad. Press, New - York .

Lematre, J., Mallet G. and Vasilescu D., (1988) Physiolog. Chem. and Phys. and Medical NMR 20 213 - 219.

Lenk, R., (1979) Chem. Phys. Letters 62 399-403.

Lenk, R., (1986) in : " Fluctuations - Diffusion and Spin Relaxation ", Elsevier / Amsterdam .

Mallet G. , Costa A., Rix-Montel M.A. and Vasilescu D., (1982) Studia Biophysica 91 167-176.

Mallet, G.,Lematre J., Leca M.and Vasilescu D. , (1987)   Applied Physics Comm. 7 57-68.

Manning, G.S., (1969) J. Chem. Phys. 51   924-933.

Manning, G.S., (1972) Biopolymers 11 937-949.

Manning, G.S., (1978) Quart. Rev. Biophys. 11 179 -

Nygaard,O.F. and Simic M.G., (1983) in : "Radioprotectors and Anticarcinogens" Nygaard,O.F. and Simic M.G. eds.) Acad. Press, New-York .

Purdie, J.W., (1979) Radiat. Res. 77 303 - 311.

Schülte - Frohlinde D, (1979) in : " Radioprotectors and Anticarcinogens " (Nygaard, O.F. and Simic M.G. eds.) Acad. Press , New-York .

Van Djik, L., Gruwel M.L.H., Jan de Bleijser W.J. and Leyte J.C., (1987) Biopolymers 26 261 - 284.

Vasilescu, D. and Mallet G., (1985) Biopolymers 24 1845 - 1850.

Vasilescu D.,Broch H. and Rix-Montel M.A. , (1986) J. of Molecular Structure Theochem. 134 367 - 380.

Yuhas J.M., (1980) in: "The Treatment of Cancer" (Sokol G.H., and Maïkel R.P. Eds.) Wiley , New - York.

WATER, IONS AND PROTEINS

THE CENTRAL HELIX of CALMODULIN and HOMOLOGS
EFFECTS of SOLVENT EXPOSURE on STABILITY

Robert H. Kretsinger, Nancy D. Moncrief, and Anthony Persechini

Department of Biology, University of Virginia
Charlottesville, VA 22901, U.S.A.

SUMMARY: The linker region of the central helix of calmodulin is unique in that it is completely exposed to solvent for two turns. It functions as a flexible tether permitting the hydrophobic patches on lobe 1,2 and on lobe 3,4 to enfold an α-helical portion of the target. A comparison of the homologs of calmodulin indicates that this flexible tether is restricted in distribution and may reflect a unique evolutionary history. We suggest that an evolutionary view of the central helix may permit chemists to focus on those factors responsible for the bend, extend functional cycle of calmodulin.

BACKGROUND

Calmodulin is a ubiquitous, calcium-modulated protein. We and others have demonstrated that the linker region of the central helix of calmodulin functions as a flexible tether. The determination that the crystal structures of troponin C and of calmodulin have long central helices was unanticipated. Then, the demonstration that this central helix bends to permit the two lobes of calmodulin to enfold its target was unprecedented.

Several homologs of calmodulin may also have central helices, some rigid, some flexible. We suggest that an evolutionary view of the central helix may permit one to focus on key experiments that reveal factors that determine the stability of an α-helix.

EF-hand proteins, structures: The EF-hand domain consists of an α-helix (E) resembling the extended forefinger of a right hand, a calcium binding loop (clenched middle finger), and second α-helix (F) (extended thumb). Most domains are demonstrated or inferred to bind one Ca^{2+} ion; some lack ligand(s) involved in calcium coordination. The crystal structures of troponin C and calmodulin (reviewed in Strynadka & James, 1989), each of which has four EF-hands, (Fig. 1a) have an overall dumbbell shape. However, contrary to prediction, the linker between domains 2 and 3 is α-helical.

In these crystal structures as well as those of parvalbumin and intestinal calcium binding protein the domains occur in pairs; all pairs nearly superimposable on one another. This observation plus numerous solution studies indicate that a pair of EF-hands is the stable functional unit.

<u>EF-hand proteins, evolution</u>: Amino acid sequences are known for more than 160 different EF-hand proteins from animals, plants, fungi, protocists, and a prokaryote. These proteins contain from two to eight EF-hand domains per monomer and most can be classified into twelve distinct subfamilies, Fig. 2. Eight individual proteins are tentatively identified as "unique"; that is, they may be the sole known representative of another subfamily (Kretsinger et al, 1988).

<u>Flexible tether of calmodulin</u>: We have proposed a model (Persechini & Kretsinger, 1989), Fig. 1b, in which the linker portion of the central helix functions as a flexible tether permitting lobes 1,2 and 3,4 to enfold the α-helical portion of the target. There are no precedents for breaking and making an α-helix as part of a binding cycle. We propose that the linker region of the central helix of calmodulin is delicately balanced between helical and bent forms.

<u>ANALYSES</u>

The predicted helicities of the central helices of various EF-hand proteins are shown in Fig. 3. The amino acid sequences are aligned for the second helix (F) of domain 2 and the first helix (E) of domain 3. The sequences of the 2,3 linker regions are aligned with gaps to accommodate the longest listed linker. Qualitatively similar profiles are obtained from the algorithms of Chou and Fasman (1977) and of Garnier et al (1978); only the former are shown. Within subfamilies the individual proteins have similar profiles. Hence, we usually present only one representative from

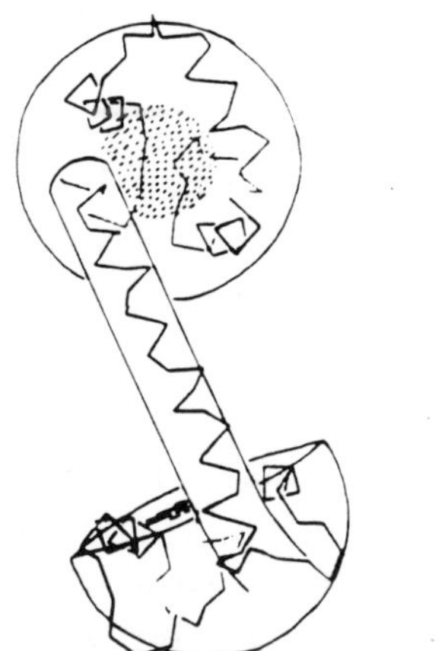

Fig. 1 α-Carbon traces of calmodulin. In 1a the extended form of the central helix consists of helix F of domain 2, the linker, and helix E3. Domains 3 and 4 are viewed down the approximate two-fold axis that relates them. The exposed hydrophobic patches on the surfaces of lobe 1,2 and lobe 3,4 are indicated by stipling. In 2a the linker is shown bent near Ser-81 permitting the two hydrophobic patches to enfold an α-helical portion of a target.

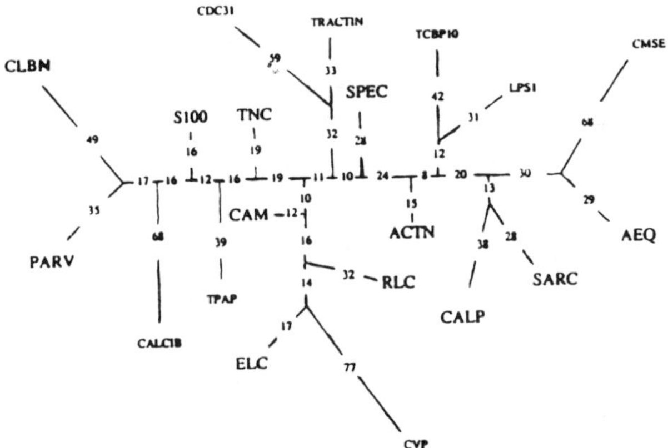

Fig. 2 PARV, parvalbumin; CLBN, calbindin; S100, S100 and other two-domain homologs, TNC, troponin C; CAM, calmodulin; ELC, essential light chain of myosin; RLC, regulatory light chain; SPEC, Strongylocentrotus purpuratus ectodermal protein; ACTN, α-actinin; AEQ, aequorin and luciferin-binding protein; CALP, calpain; and SARC, sarcoplasmic calcium-binding protein. CALCIB, calcineurin B from Bos; TPAP, troponin C from Astacus; CVP, calcium vector protein from Branchiostoma; TRACTIN, caltractin from Chlamydomonas; CDC31, cdc31 gene product from Saccharomyces; TCBP10, 10 kD calcium-binding protein from Tetrahymena, LPS1, eight-domain protein from Lytechinus; and CMSE, calcium-binding protein from Streptomyces.

each subfamily; if groups within subfamilies are different, a representative from each group is shown.

For both human CAM and human skeletal TNC a turn, and corresponding very low probability of helix formation, is predicted within the linker; while, the F2 and E3 regions show moderate to strong tendency to form helices. Even though the CAM of Saccharomyces is inferred not to bind calcium in domain 4 and cardiac TNC not to bind calcium in domain 1, they have helix prediction profiles similar to those of calmodulin and are strongly inferred to have the same dumbbell structures in solution.

Cal-1 and squidulin are both members of the CAM subfamily; both the nematode and the squid also contain CAM itself. Cal-1 shows less tendency to bend than does CAM while squidulin contain two Pro's in its linker.

The TNC from the tunicate branches near the base of the subfamily. Its linker is four residues longer than vertebrate TNC and contains a Pro. In contrast the TPAP formally lies outside the subfamily, see Fig. 2; yet, its linker is eleven residues long and shows the highest probability of helix formation of any known linker. It binds calcium in domains 2 and 4.

The RLC's contain a Pro in their linkers. Even though they function by binding calcium in their first domains, they probably do not extend and bend during their functional cycles.

The ELC's of animals all contain Pro in helix F2 and none bind calcium. All are predicted to have non-helical linkers as well. The sites of the bends vary and in the fruit fly the bend at the C-end is complemented by a strong helical probability at the N-end of the linker. The light chain from the slime mold clusters with the ELC's; however, the designation "ELC" may be inconsistent. It binds two Ca^{2+} ions and has a Pro in the linker, as do the RLC's.

CVP from amphioxus has been inferred to interact with target proteins in a manner similar to CAM (Cox, 1989). Its seven residue linker and helix F2 are predicted to be nonhelical.

CALCIB is an integral subunit of a heterodimeric protein phosphatase. Its helix F2 as well as its eight residue linker is predicted to be nonhelical. One might not expect it, as a subunit, to undergo the bend, extend cycle.

Both the six domain CLBN and the four domain fragment from calretinin contain Pro's in their helices F2, with strong tendency to helix break in the linker of calretinin. The functions of these proteins are unknown; the sixteen residue linker seems especially precarious as a candidate for helicity.

Although CDC31 from a yeast and TRACTIN from a protocist are both tentatively classified as unique, they may both ultimately be seen as members of a single subfamily. The function of CDC31 is unknown; TRACTIN is associated with spindle bodies. Both bind four Ca^{2+} ions; both have linkers eight residues long. A helix break is predicted in TRACTIN at the position of a Pro in CDC31.

The functions of the SPEC's, remain unknown; although, all four domains are inferred to bind calcium. Isoform 2a is strongly predicted to have a bend in the linker region; the strongest tendency to bend in isoform 1 is in the N-end of helix E3. The fact that "1" has a nine residue linker and "2a" has a six residue linker may reflect different functions.

LPS1 contains eight domains. Both the 2,3 and the 6,7 linkers are predicted to be nonhelical. The three most distantly related subfamilies and the calcium binding protein from the prokaryote, CMSE, all have one or two Pro's in their linkers.

AEQ is a calcium-activated luciferase; part of the 23 residue "linker" may be involved in forming the active site. Two of the four EF-hands at the C-terminus of the CALP's lend calcium sensitivity to this Cys-protease. Neither targets nor functions have been found for CMSE or for the SARC's.

DISCUSSION

Our analyses have generated many more questions than answers:

1. Is the flexible tether unique to calmodulin or do the central helices of other four (or more) domain homologs bend and extend as part of their functional cycles? Beyond calmodulin and troponin C we cannot predict with confidence the existence of other central helices. We would be even more cautious in suggesting additional flexible tethers. This bend, extend cycle may be unique

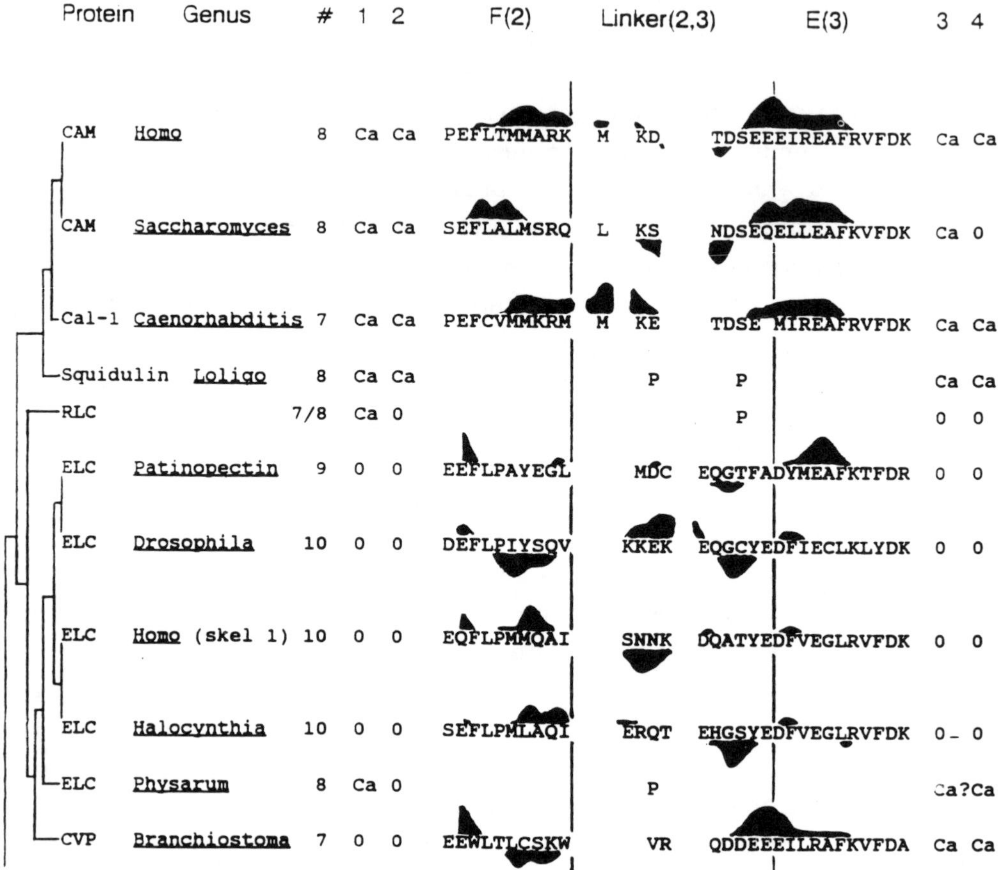

Fig. 3   The residues of the linker (2,3) are listed as properly aligned with linkers of other subfamilies; the number of linker residues are listed under "#".   In the columns -- 1, 2, 3, and 4 -- is indicated whether the domains are demonstrated or inferred to bind calcium (Ca) or not (0).   The symbol "Ca?Ca" with *Physarum* ELC indicates that both domains 3 and 4 might bind calcium as judged by the sequence; "CaCa/0" with SARC indicates that different members are inferred to bind calcium with either or both domains 3 and 4.   Sequences are not listed for homologs that contain a Pro in their linkers.   Since a strong structural similarity is inferred for the proteins within a subfamily, if one member contains a Pro, then all members are inferred to be non-helical at that position.   This does not preclude the possibility of a helix preceding and following the Pro.   The branches and nodes to the left of the protein names show their relationshps as indicated in Fig. 2.   Acronyms are defined in the legend to Fig. 2; *Chlamydo.* = *Chlamydomonas*, *Strongylo.* = *Strongylocentrotus*, *Lytechin.* = *Lytechinus*.   Two-domain homologs (S100, ACTN, TCBP10) and the three-domain PARV's, were not included in this analysis.

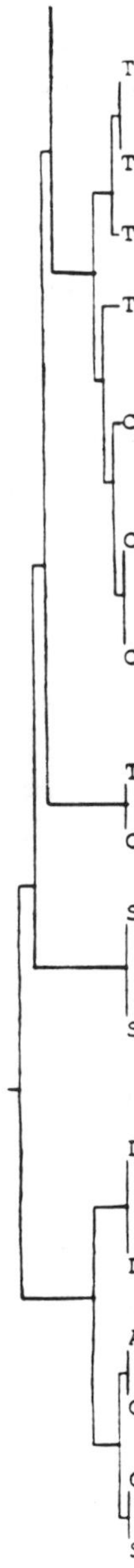

Protein	Genus	#	1	2	F(2)	Linker(2,3)	E(3)	3	4
TNC	Homo (skel)	11	Ca	Ca	EEFLVMMVRQ	MKEDAK	GKSEEELAECFRIFDR	Ca	Ca
TNC	Homo (card)	11	0	Ca	DEFLVMMVRC	MKDDSK	GKSEEELSDLFRMFDK	Ca	Ca
TNC	Halocynthia	15	0	Ca			P	0	Ca
TPAP	Astacus (2)	11	0	Ca	EEFVELAAKF	LIEEDE	EALKAELKEAFRIYDK	0	Ca
CALCIB	Bos	8	Ca	Ca	KEFIEGVSQF	SVK	GDKEQKLRFAFRIYDM	Ca	Ca
Calretinin	Gallus	15	Ca	0	AELAQILPTEENFLLCFR	QHVGSSSEFMEAWRRYDT		Ca	Ca
CLBN	Homo	16	Ca	0	VELAHVLPTEENFLLLFRCQQLKSCEEFMKTWRKYDT			Ca	Ca
TRACTIN	Chlamydo.	8	Ca	Ca	EEFLTMMTAK	M GE	RDSREEILKAFRLFDD	Ca	Ca
CDC31	Saccharomyces	8	Ca	Ca			P	Ca	Ca
SPEC 1	Strongyl.	9	Ca	Ca	SEMLMGIAEQ	M VK	WTWKEEHYTKAFDDMDK	Ca	Ca
SPEC 2a	Strongyl.	6	Ca	Ca	SEFLMRKALQ	W RG	REVQLTKAFDDLDK	Ca	Ca
LPS1(1-4)	Lytechin.	9	Ca	Ca	DEFILYM EG	S TK	ERLYSSEIKQMFDDLDK	Ca	Ca
LPS1(5-8)		4	Ca	Ca	NEFFMHL DG	V SK	DHIKQQFMAIDK	Ca	0
AEQ	Aequorea	23	Ca	0		P		Ca	Ca
CMSE	Streptomyces	15	Ca	Ca			P	Ca	Ca
CALP		6	Ca	Ca			PP	0	0
SARC		15	Ca	Ca/0		P	P	CaCa/0	

to the calmodulins.

    2.   What is the source of the unanticipated stabilization? Sundaralingam et al (1985) proposed a complex network of stabilizing hydrogen bonds between residues three or four removed along the helix, but few of these are seen in the crystal structures.

    3.   To what extent and how do the two lobes communicate with one another? At least three models -- dynamic equilibrium, target creeping, and helix transmission -- have been considered. Each reasonably assumes a difference in conformation and in energy between the calcium-bound and apo-forms. At this time there is little evidence to support or to refute any of the three models. There have been, however, indications of some sort of communication between lobes (Tsalkova & Privalov, 1985).

CONCLUSION

Calmodulin and its homologs are a very promising system for exploring the stability of the α-helix. Obviously, quite sophisticated chemistry will be required to understand how energy and information is transmitted along the linker helix. This chemistry might well be guided by considering the diverse lobes and linkers already employed in nature.

REFERENCES

Chou, Y., and Fasman, G.D. (1977) J. Mol. Biol. 115, 135-175.

Cox, J.A. (1990) In: Stimulus-response Coupling: The Role of Intracellular Calcium" (J.R. Dedman and V.L. Smith, Eds), Tolford Press.

Garnier, J., Osguthrope, D.J. and Robson, B. (1978) J. Mol. Biol. 120, 97-120.

Kretsinger, R.H., Moncrief, N.D., Goodman, M. and Czelusniak, J. (1988) In: The Calcium Channel: Structure, Function and Implications (M. Morad, W.G. Naylor, S. Kazda, and M. Schramm, Eds), Springer Verlag, pp. 16-35.

Persechini, A. and Kretsinger, R.H. (1988) J. Biol. Chem. 263, 12175-12178.

Sundaralingam, M., Rao, S.T., Drendel, W. and Greaser, M.L. (1985) Int. J. Quant. Chem. 12, 153-160.

Strynadka, N.C.J. and James, M.N.G. (1989) Annu. Rev. Biochem. 58, 951-998.

Tsalkova, T.N. and Privalov, P.L. (1985) J. Mol. Biol. 181, 533-544.

CONTROL OF PROTEIN KINASE C FUNCTION: AN INHIBITOR STUDY

C.E. Hensey, D. Boscoboinik and A. Azzi

Institut für Biochemie und Molekularbiologie, Universität Bern,
Bühlstrasse 28, CH-3012 Bern, Switzerland

SUMMARY: The effect of various inhibitors on the activity of
protein kinase C (PKC) was investigated. Studies in vitro with
purified rat brain PKC were carried out alongside studies using
tumor and normal tissue cell lines. Suramin, an anti-cancer
drug, was found to inhibit PKC activity and induce differentiation in neuroblastoma cell clone NB2A. Staurosporine, a potent
inhibitor of PKC activity, also inhibited the phorbol ester
induced activation of the Na^+/H^+ antiporter in smooth muscle
cells. However once the antiporter had been activated,
inhibitors of PKC were not effective supporting a mode in which
the Na^+/H^+ antiporter conserves memory of its activation.

INTRODUCTION

The Ca^{2+} and phospholipid dependent kinase, protein kinase C
(PKC), first described by Nishizuka as a protease-activated
kinase (Takai et al., 1977; Inoue et al., 1977) is thought to
play an important role in signal transduction across the cell
membrane (for review see Nishizuka 1984, 1986). The enzyme has
been implicated in numerous other biological processes including
tumor promotion, differentiation, secretion, neural synaptic
transmission, muscle contraction, platelet aggregation and
membrane channel and transporter modulation (Nishizuka, 1986;
Liles et al., 1986; Witters et al., 1986; Froscio et al., 1988).

PKC with an apparent molecular wt. (Mr) of 84000 Da is unique in that it requires both Ca^{2+} and phospholipid for activation. An important feature of the enzyme is its activation by the tumor promoters phorbol esters (e.g. phorbol 12-13 dibutyrate) and by diacylglycerol. Both these substances act by lowering the Ca^{2+} required for activation (Takai et al., 1979; Castagna et al., 1982) and phorbol 12-13 dibutyrate/ diacylglycerol competition studies suggest that both compounds interact with the same or adjacent binding sites on the enzyme phospholipid complex (Sharkey et al., 1984; Sharkey & Blumberg, 1985). Two major structural domains have been described for the enzyme, the phospholipid/diacylglycerol/ phorbol ester binding regulatory domain ($\approx$ 30 KDa) and the ATP/substrate binding catalytic domain ($\approx$ 50 KDa) were predicted from the deduced amino acid sequences (Parker et al., 1986; Coussens et al., 1986). PKC has been shown to be an increasingly large family of gene products and closely related isozymes encoded by at least six distinct genes have been identified (Parker et al., 1986; Coussens et al., 1986). Heterogeneity of cellular localization and function has also been reported (Huang et al., 1987; Ohno et al., 1987; Nishizuka, 1988).

The importance of PKC in the control of many cellular functions implies that the study of inhibitors of PKC is important for the understanding of physiological control mechanisms and drug action. We have investigated the inhibition of PKC by both physiological and non-physiological compounds. Auranofin, a gold (I) compound currently used as an antirheumatic agent, and vitamin E were shown to inhibit PKC activity (Mahoney et al., 1989; Mahoney & Azzi, 1988).

Suramin, a polysulphonated naphthylurea, was the first widely accepted antiparasitic drug to be developed and is still one of the most commonly prescribed antitrypanosomal drugs. The inhibition of protein kinases in <u>Trypanosomatida</u> by suramin has been demonstrated (Walter, 1980; Misset & Opperdoes, 1987). In recent studies suramin was used in the chemotherapy of acquired immune deficiency syndrome (AIDS) (Collins et al., 1986) but host toxicity precluded its usefulness. The drug is currently under investigation for the treatment of advanced malignancy and has exhibited antitumor activity in a number of systems (Stein et al., 1989). Since suramin has been shown to interact with other kinases we have considered and tested the possibility that suramin antitumor activity is due to its inhibition of PKC.

Most mitogens involved in the regulation of growth of mammalian cells trigger a series of events including the activation of PKC and intracellular alkalinization mediated by the Na^+/H^+ exchanger (Moolenaar et al., 1984; Besterman & Cuatrecasas, 1984; Rozengurt, 1986). Stimulation of the amiloride-sensitive Na^+/H^+ antiporter is one of the earliest and nearly universal responses of quiescent cells to growth-promoting agents such as phorbol esters (Berk et al., 1987). This activation results in the extrusion of protons and the concomitant alkalinization of the cytoplasm.

Previous studies have demonstrated the existence of two pathways for regulation of the Na^+/H^+ antiporter activity: one which involves the inositol triphosphate stimulated increase in cytosolic free Ca^{2+} concentration and another mediated by the activity of PKC (Moolenaar et al., 1984; Vinge et al., 1985; Grinstein et al., 1985; Ives & Daniel, 1987). There are some lines of evidence indicating that the activation of Na^+/H^+ exchanger by phorbol esters is mediated by stimulation of PKC (for reviews, see Moolenaar, 1988; Grinstein et al., 1989).

Staurosporine, a microbial alkaloid produced by _Streptomyces actuosus_ (Omura et al., 1977) is one of the most potent inhibitors of PKC (Tamaoki et al., 1986). By using such an inhibitor we investigated whether inhibition of PKC prevented the mitogen-induced responses, such as cytoplasmic alkalinization, and the dynamics of activation of the Na^+/H^+ antiporter in smooth muscle cells.

RESULTS AND DISCUSSION

The inhibition of purified rat brain PKC by suramin is shown in Fig. 1.

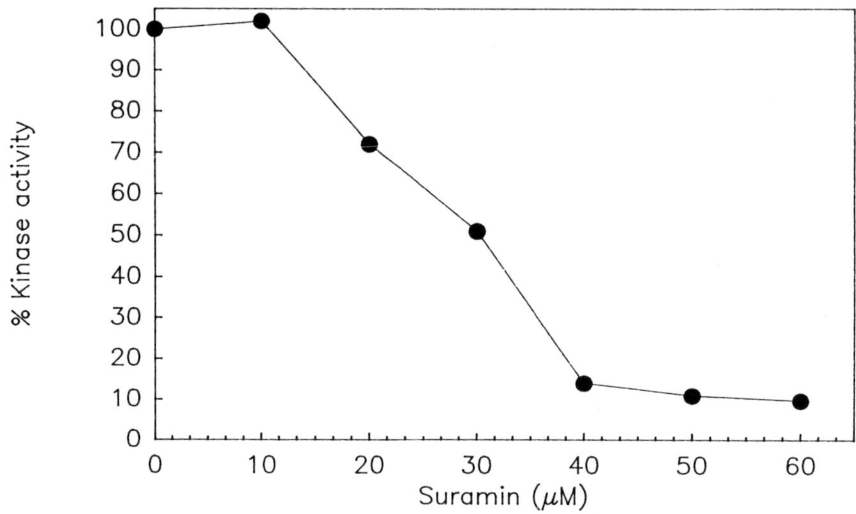

Fig. 1. Concentration dependent inhibition of PKC activity by suramin. Results are the mean of triplicate experiments.

A half-inhibitory concentration of 30 μM, for the assay conditions employed, was calculated. The lack of inhibition, and sometimes the slight activation, of the enzyme at concentrations below 10 μM is, at present, not clear. A kinetic analysis of the inhibition of PKC by suramin is shown in Fig. 2. The type of inhibition appears to be competitive with respect to ATP (K_i=10 μM). The data points were not fitted optimally by a straight line, suggesting the existence of heterogeneity in the preparation. Such a phenomenon may be consistent with a different degree of inhibition by suramin of the different isoforms of the enzyme. The lack of inhibition at low concentrations of suramin may be related to the same phenomenon. The inhibition of the Na^+-K^+ ATPase (Fortes et al., 1973) and Ca-ATPase (Layton & Azzi, 1974) by suramin, together with the present results are consistent with previous kinetic and computer modelling studies of protein kinases from <u>Trypanosomatida.</u>

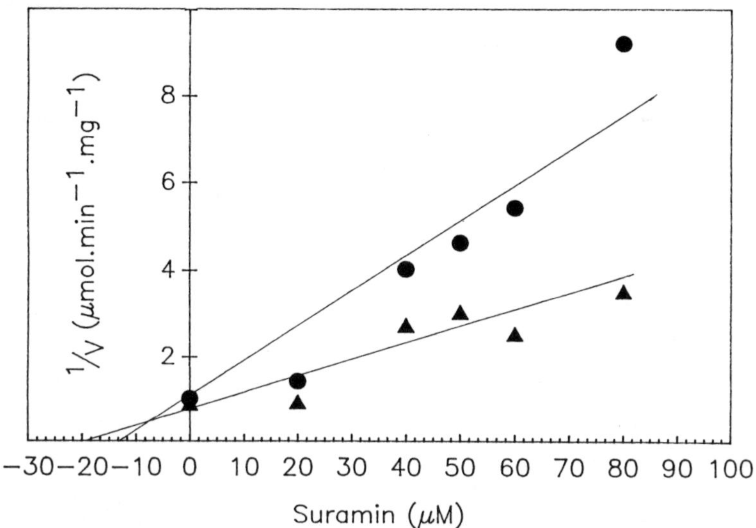

Fig. 2. Dixon plot of the inhibition of PKC by suramin. PKC activity was measured in the presence of different concentrations of suramin at two ATP concentrations, 10 μM (●) and 40 μM (▲). Results are the mean of triplicate experiments.

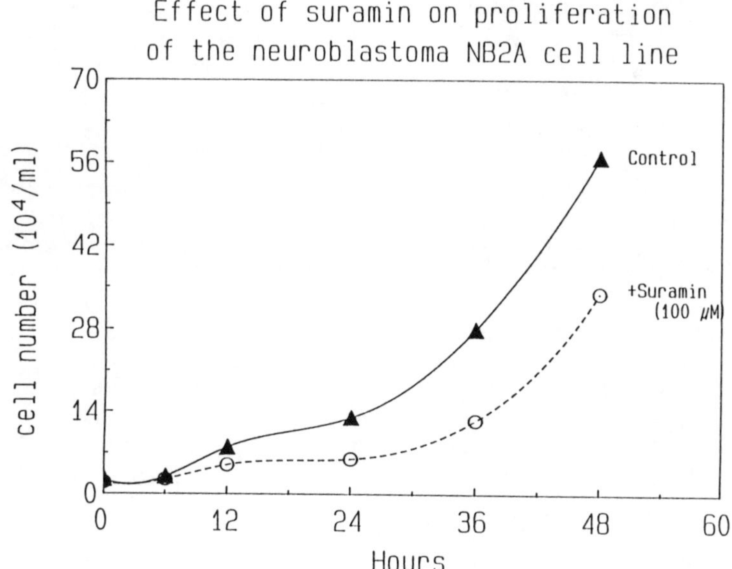

Fig. 3. Effect of suramin on serum-induced proliferation of NB2A cells in culture. The quiescent cultures were stimulated by changing the medium to one containing 5% FCS, with (○) or without (▲) suramin (100 μM).

In these studies suramin was shown to inhibit protein kinase I by competing for the ATP binding site (Walter, 1980) and computer modelling of phosphoglycerate kinase have shown how suramin could be fitted to the ATP binding site and thereby inhibit kinase activity (Hart et al., 1989).

To investigate the effect of suramin on intact cells, quiescent cultures of neuroblastoma cells (clone NB2A) maintained in 0.1% FCS were stimulated to growth by the addition of 5% FCS to the culture medium. In untreated cultures an increase of 20-fold in the cell number was observed after 48 hours incubation. The presence of 100 μM suramin in the culture medium produced a diminution of approximately 50% in the rate of cell multiplication in the first 30 hours. Subsequently, however, the rate of growth of the cells, in presence and absence of suramin, became equal (Fig. 3). The conclusion which can be drawn from this experiment is that suramin slows down, at

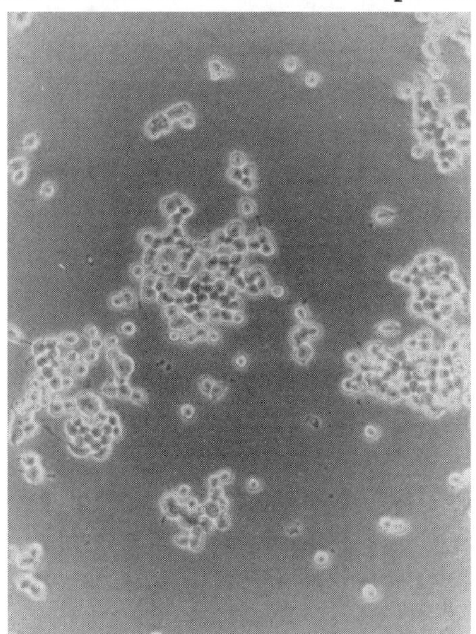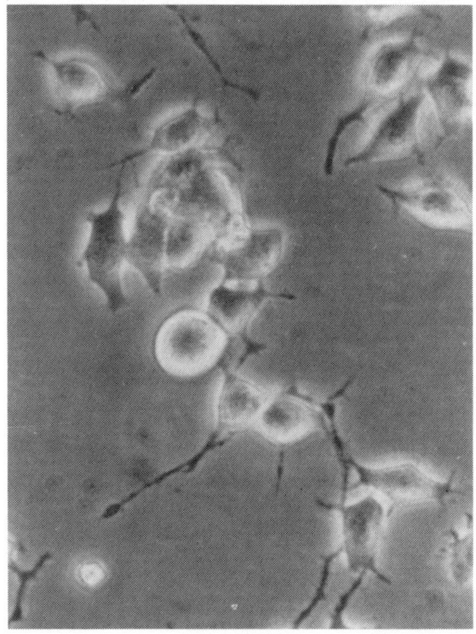

Fig. 4. Effect of suramin on the morphology of NB2A neuroblastoma cells. Cells were plated at 1 x 10^5/well in 6-well plates and suramin was added 24 h later in the presence of 2% FCS. After 48 h of treatment, the degree of differentiation, as evidenced by the enlargement of neurites, is observed in suramin-treated cells (right), but not in control cells (left).

least in the first 30 hours of incubation, the process of cell multiplication. Why the effect is not prolonged with time is not, at present, clear. What is, however, evident is that at 48 hours of the treatment the cells have changed rather dramatically their morphology. The spherical and generally separated cells emit neurites, which in many cases are branched, indicating a certain degree of differentiation. The cells acquire a polygonal shape and tend to associate more frequently in lumps (Fig. 4). Whether suramin produces the above described morphological changes by solely interacting with PKC or by multiple interactions with different proteins (possibly utilizing ATP as a substrate) remains to be elucidated. The analogous findings obtained with staurosporine a potent PKC inhibitor, suggest a central role of PKC in the effects of suramin at a cellular level. The newly described inhibition of PKC through the interaction of suramin with the catalytic domain of the enzyme suggests that the inhibition of growth, and induction of differentiation in neuroblastoma cells, could be mediated through the inhibition of PKC. However, other mechanisms by which this drug exhibits anti tumor activity have been reported. Suramin was shown to block the binding of a range of tumor growth factors to their cell surface receptors thereby inhibiting tumor growth (Coffey et al., 1987) and another report suggests the antitumor effect may be due to the drug causing an increase in tissue glycosaminoglycans which have important effects in tumor biology including differentiation (Stein et al., 1989). However, since suramin is an inhibitor of PKC we see this as an alternative reason for suramin antitumor activity.

In Fig. 5 the intracellular alkalinization produced by the addition of PMA to smooth muscle cells suspended in HCO_3-free saline solution, is followed by using the pH-sensitive fluorescent dye BCECF. Intracellular pH was monitored for 15 min after the addition of 0.2 μM PMA. The treatment with PMA produced, after an initial lag period of one minute, a sustained alkalinization of approximately 0.12 pH units during the first 10 min of incubation (upper curve, Fig. 5) The PMA-induced alka-

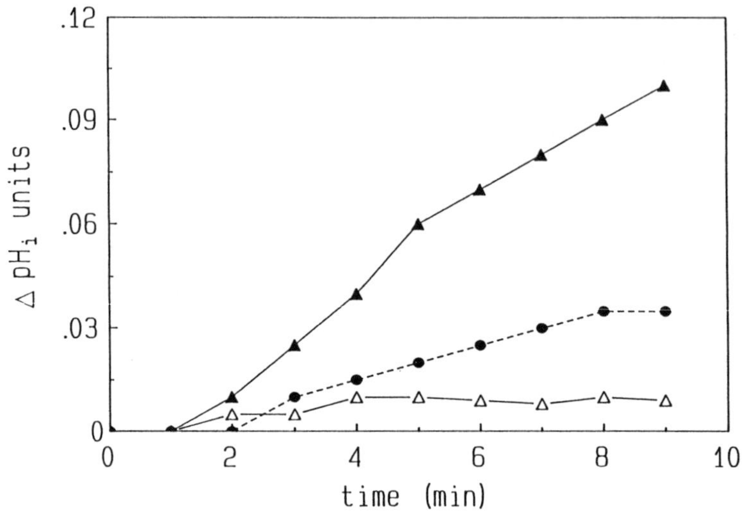

Fig 5. The resting pHi of A7r5 cells incubated in HCO3-free media was 7.05 + 0.05 (n=5). Addition of PMA (0.2 μM) provoked a cytosolic alkalinization ($\blacktriangle$), which was blocked by 100 μM amiloride ($\triangle$) or 20 nM staurosporine ($\bullet$).

linization was inhibited more than 90% by the addition of 100 μM amiloride when both drugs were added together (lower curve, Fig. 5), indicating that the alkalinization was mediated by the activation of the Na^+/H^+ exchanger. Phorbol esters have been shown to increase the activity of the Na^+/H^+ antiporter in many cell lines (Moolenaar et al., 1984; Vinge et al., 1985; Berk et al., 1987). It is postulated that the activation is related to phosphorylation of the exchanger by PKC (Grinstein & Rothstein, 1986; Cassel et al., 1986; Owen et al., 1989), thus the delay of one minute in the onset of alkalinization may be due to a threshold of phosphorylation which should be reached before the exchanger becomes activated.

The effect of staurosporine on the PMA-induced cell alkalinization mediated by the Na^+/H^+ exchanger is represented by the middle curve of Fig. 5. Staurosporine, a potent inhibitor of PKC, also caused a marked reduction in the alkalinization of A7r5 cells, although not as potent as that produced by amiloride.

PKC independent activation of the Na^+/H^+ exchanger has been described in several systems (Magnaldo et al., 1986; Hatori et al., 1987; Little et al., 1988). However, the observation that PMA induced the activation of the Na^+/H^+ exchanger (Fig. 5) and that staurosporine inhibited this PMA-induced alkalinization in a dose-dependent manner (Fig. 6) infers the involvement of PKC in the stimulation of the exchanger in A7r5 cells. Staurosporine at a concentration of 4-5 nM produced a 50% inhibition of the PMA-induced cytosolic alkalinization.

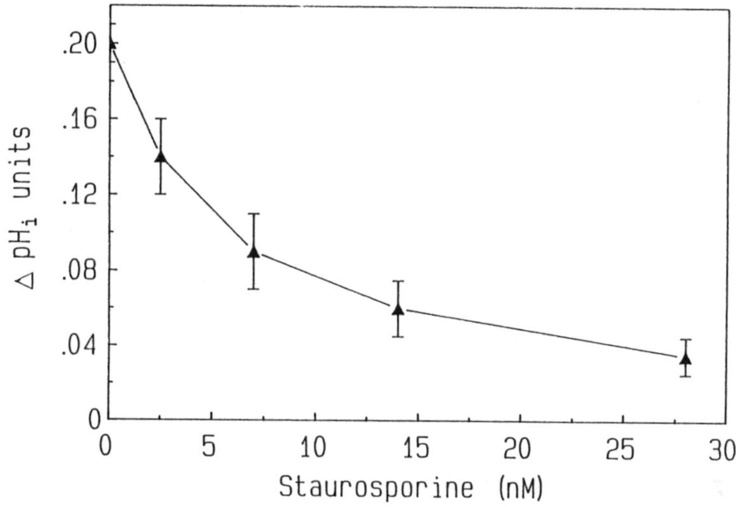

Fig 6. Effect of staurosporine concentration on the PMA-induced changes in pHi in A7r5 cells.The cells were preincubated for 5 min with staurosporine and then 0.2 μM PMA was added. After 15 min, the changes in pHi for each concentration of staurosporine were measured.

Regulation of an effector protein by phosphorylation necessarily requires protein phosphatases in addition to protein kinases. The concerted action of the kinase/phosphatase pair would determine the actual level of activation of the protein. It is possible that the Na^+/H^+ exchanger is a multiply phosphorylated protein that is activated to a different extent depending on its phosphorylation state (Cassel et al., 1986; Owen et al., 1989). In order to determine how stable in time the

activation of the exchanger was after its phosphorylation was
blocked, the experiment depicted in Fig. 7 was performed   under

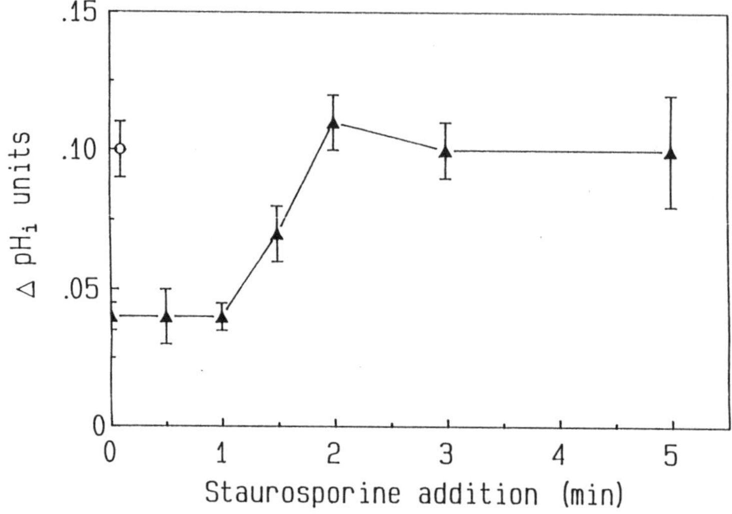

Fig 7. A7r5 cells were stimulated with 0.2 μM PMA and
staurosporine (5 nM) was added at the indicated times after
stimulation. The change in pHi at 7 min after the induction by
the phorbol ester was measured in the absence ($\bigcirc$) or in the
presence of staurosporine ($\blacktriangle$).

the same conditions as in Fig. 5 except that staurosporine was
added at different times after stimulation by PMA. Addition of
PMA to A7r5 cells at time 0 induced an alkalinization of about
0.1 pH units in 7 min. Consistently, staurosporine inhibited up
to 75% when added together with PMA. The same inhibition was
observed if staurosporine was added 0.5 or 1 min after the
addition of PMA, indicating that during this time the antiporter
was still sensitive to the effects of the inhibitor. However, as
soon as total activation of the antiporter had been achieved the
effect of staurosporine became less potent (addition at 1.5 min)
or irrelevant (addition at 2 min and thereafter). It should be
noticed that after addition of PMA (Fig. 5), a one minute lag
was observed before the onset of alkalinization.

An obvious control was to show that PKC itself did not,
after activation, become refractory to staurosporine inhibition

(Fig. 8). Using purified rat brain PKC phosphorylation of histone III-S by PMA-activated PKC proceeded linearly with time up to 8 min (upper curve) and this activity was inhibited almost

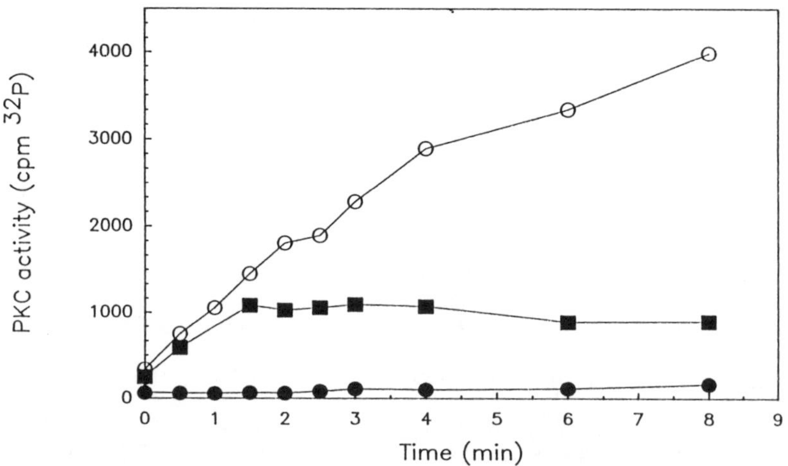

Fig.8. Inhibition of PKC by staurosporine. PKC activity was measured at different time points in the absence (O) and in the presence of staurosporine (● , + staurosporine at zero time, ■, + staurosporine at 1 min).

100% by staurosporine (lower curve). Addition of staurosporine 1 min after PKC activation by PMA gave immediate and full inhibition of the kinase activity for the duration of the reaction. It can be therefore concluded that PKC can be inhibited before or after its activation by PMA and the results shown in Fig. 7 cannot be interpreted in terms of a lack of inhibition of PKC developing after PMA activation of the enzyme. Thus the conclusion remains that the Na^+/H^+ antiporter remains active despite PKC inhibition.

The above conclusion cannot be generalized to all PKC substrates. Activation of the respiratory burst in neutrophils by PMA can be inhibited at any time after stimulation of O^{2-} production (Mahoney et al., 1989). The conclusion that the Na^+/H^+ antiporter has a rather long "memory" of its own activation has some bearing on phenomena of great biological importance such as post synaptic potentiation and tumor cell differentiation.

Acknowledgements: The support of C.E. Hensey by Schweizerische Krebsliga is gratefully acknowledged.
We wish to thank Prof. Michel Lazdunski for the generous gift of the A7r5 cell line.

REFERENCESTS

Berk, B.C., Aronow, M.S., Brock, T.A., Cragoe, E., Gimbrone, M.A., and Alexander, R.W. (1987) J. Biol. Chem. 262, 5057-5064.
Besterman, J.M., and Cuatrecasas, P. (1984) J. Cell. Biol. 99, 340-343.
Cassel, D., Katz, M., and Rotman, M. (1986) J. Biol. Chem. 261, 5460-5466.
Castagna, M., Takai, Y., Kaibuchi, K., Sano, K., Kikkawa, U., and Nishizuka, Y. (1982) J. Biol. Chem. 257, 7847-7851.
Coffey, R., Leof, E., and Shipley, G. (1987) J. Cell. Phys. 132, 143-148.
Collins, J.M., Klecker, R.W., Yarchoan, R., Lane, H.C., Fauci, A.S., Redfield, R.R., Broder, S.B., and Myers, C.E. (1986) J. Clin. Pharmacol. 26, 22-26.
Coussens, L., Parker, P.J., Rhee, L., Yang Feng, T.L., Chen, E., Waterfield, M.D., Francke, U., and Ullrich, A. (1986) Science 233, 859-866.
Fortes, P.A., Ellory, J.C., and Lew, V.L. (1973) Biochim. Biophys. Acta. 318, 262-272.
Froscio, M., Solanki, V., Murray, A.W., and Hurst, N.P. (1988) Biochem. Pharmacol. 37, 366-368.
Grinstein, S., Cohen, S., Goetz, J.D., Rothstein, A., and Gelfand, E.W. (1985) Proc. Natl. Acad. Sci U.S.A. 82, 1429-1433.
Grinstein, S., and Rothstein, A. (1986) J. Membr. Biol. 90, 1-12.
Grinstein, S., Rotin, D., and Mason, M.J. (1989) Biochim. Biophys. Acta 988, 73-97.
Hart, D., Langridge, A., Barlow, D., and Sutton, B. (1989) Parasitology Today 5, 117-120.
Hatori, N., Fine, B.P., Nakamura, A., Cragoe, E.Jr., and Aviv, A. (1987) J. Biol. Chem. 262, 5073-5078.
Huang, F.L., Yoshida, Y., Nakabayashi, H., and Huang, K.P. (1987) J. Biol. Chem. 262, 15714-15720.
Inoue, M., Kishimoto A., Takai, Y., and Nishizuka, Y. (1977) J. Biol. Chem. 252, 7610-7616.
Ives, H.E., and Daniel, T.O. (1987) Proc. Natl. Acad. Sci. U.S.A. 84, 1950-1954.
Layton, D., and Azzi, A. (1974) Biochem. Biophys. Res. Commun. 59, 322-325.

Liles, W.C., Hunter, D.D., Meier, K.E., and Nathanson, N.M. (1986) J. Biol. Chem. 261, 5307-5313.

Little, P.J., Weissberg, L., Cragoe, E.J., Jr., and Bobik, A. (1988) J.Biol.Chem. 263, 16780-16786.

Magnaldo, I., L'Allemain, G., Chambard, J.C., Moenner, M., Barritault, D., and Pouyssegur, J. (1986) J. Biol. Chem. 261, 16916-16922.

Mahoney, C.W., Azzi, A. (1988) Biochem. Biophys. Res. Commun. 154, 694-697.

Mahoney, C.W., Hensey, C.E., Azzi, A. (1989) Biochemical Pharmacology 38, 3383-3386.

Mahoney, C.W., and Azzi, A. (1989) In: Molecular Biology of Human Diseases, Vol 1 (J.W. Gorrod, O. Albano and S. Papa, Eds), Ellis Horwood Ltd, England, pp. 33-46.

Misset, O., and Opperdoes, F.R. (1987) Eur. J. Biochem. 162, 493-500.

Moolenaar, W.H., Tertoolen, L.G., and de Laat, S.W. (1984) Nature 312, 371-374.

Moolenaar, W.H. (1988) Annu. Rev. Physiol. 48, 363-376.

Nishizuka, Y. (1984) Nature 308, 693-698.

Nishizuka, Y. (1986) Science 233, 305-312.

Nishizuka, Y. (1988) Nature 334, 661-665.

Ohno, S., Kawasaki, H., Imajoh, S., Suzuki, K., Inagaki, M., Yokokura, H., Sakoh, T., and Hidaka, H. (1987) Nature 325, 161-166.

Omura, S., Iwai, Y., Hirano, A., Nakagawa,A., Awaya, J., Tsuchiya, H., Takahashi, Y., and Masuma, R. (1977) J. Antibiot. 37, 275-282.

Owen, N.E., Knapik, J., Strebel, F., Tarpley, W.G., and Gorman, R.R. (1989) Am. J. Physiol. 256, C756-C763.

Parker, P.J., Coussens, L., Totty, N., Rhee, L., Young, S., Chen, E., and Stabel, S. (1986) Science 233, 853-859.

Rozengurt, E. (1986) Science 234, 161-166.

Sharkey, N.A., Leach, K.L., and Blumberg, P.M. (1984) Proc. Natl. Acad. Sci. U.S.A. 81, 607-610.

Sharkey, N.A., and Blumberg, P.M. (1985) Biochem. Biophys. Res. Commun. 133, 1051-1056.

Stein, C.A., LaRocca, R.V., McAtee, T.N., and Myers, C.E. (1989) J. Clin. Oncol. 7, 499-508.

Takai,.Y., Kishimoto, A., Inoue, M., and Nishizuka, Y. (1977) J. Biol. Chem. 252, 7603-7609.

Takai, Y., Kishimoto, A., Kikkawa, U., Mori, T., and Nishizuka, Y. (1979) Biochem. Biophys. Res. Commun. 91, 1218-1224.

Tamaoki, T., Nomoto, H., Takahashi, I., Kato, Y., Morimoto, M., and Tomita, F. (1986) Biochem. Biophys. Res. Commun. 135, 397-402.

Vigne, P., Frelin, C., and Lazdunski, M. (1985) J.Biol.Chem. 260, 8008-8013.

Walter, R.D. (1980) Molec. Biochem. Parasitol. 1, 139-142.

Witters, L.A., Vater, C.A., and Lienhard, G.E. (1985) Nature 315, 777-778.

Water and Ions in Biomolecular Systems
Advances in Life Sciences
© 1990 Birkhäuser Verlag Basel

ON THE STRUCTURE AND ELASTICITY OF ELASTIN

A.M. Tamburro and V. Guantieri

Departments of Chemistry, Universities of Basilicata and Padova, Italy

SUMMARY: peptides and polypeptides comprising the sequence X-Gly-Gly (X=Val, Ala, Leu) have been synthesized by classical methods. They represent either analogs or fragments of the prolyl devoid sequences of elastin. The conformation of the peptides has been studied in a wide range of conditions by using CD and mono- and two-dimensional proton NMR. The obtained results are discussed in terms of the structure and elasticity of elastin.

INTRODUCTION

The current perspectives of the nature of elastin elasticity are essentially represented by two alternative structural models:

i) a network of random chains within the elastin fibers, according to the classical theory of elasticity (Flory, 1969). An ensemble of such chains with a random distribution of end-to-end chain lengths contitutes the source of the entropic elasticity of the protein;

ii) at variance with the classical theory, Urry (1988 a,b) proposed that the entropic elasticity of elastin derives from regular spiral (ß-spiral) structures, with fixed end-to-end chain lengths, in which internal librational motion provides

the entropic elastomeric force. According to studies carried
out on sinthetic poly(Val-Pro-Gly-Val-Gly), a repetitive
pentapeptide sequence present in elastin, the ß-spirals are
characterized by type II ß-turns acting as spacers between
the turns of the spiral. Between the ß-turns are suspended
dipeptide segments within which large-amplitude low-frequency
librations can occur. The decrease in amplitude of the librations on extension constitutes a large decrease in the entropy of the segment, providing the driving force for return to
the relaxed state. The actual molecular conformation would
strongly depend on temperature. In fact, according to Urry,
an inverse temperature transition occurs on raising the
temperature from 20° to 40° and this should induce the relatively disordered, largely extended series of ß-turns to wind
up into an elastic ß-spiral structure, in which the intramolecular hydrophobic interactions are optimized.
In the past years our laboratory has extensively investigated the physico-chemical properties of soluble elastin as
well as of synthetic polypeptides related to elastin (see the
followings). As early as 1973 (Abatangelo et al., 1973) we
showed   ß-turn conformations in peptides obtained by alkali
ne hydrolysis of elastin. Later, the conformation of α-ela-
stin (a product derived from the oxalic acid hydrolysis of
elastin) was studied (Tamburro et al., 1977; Tamburro et al.
1978; Guantieri et al., 1987). The results clearly indicated
that the ability of the polypeptide chains to adopt the ß-
-turn structure was strictly related to the environment. In
particular, the stability of the turns was shown to increase
by decreasing the water activity.
More recently, it has become clear that repetitive sequences
are abundant in the amino acid sequence of elastin (Raju and

Anwar, 1987; Bressan et al., 1987; Indik et al;, 1987).
As a matter of fact, two kinds of repetitive sequences are
essentially present in elastins (Fig.1).

Gly-X-Pro: (Val-Pro-Gly), (Val-Pro-Gly-Gly), (Val-Pro-Gly-
              Val-Gly), (Ala-Pro-Gly-Val-Gly-Val)
Gly-X-Gly: (Gly-Val-Gly-Gly-Leu), (Gly-Leu-Gly-Gly-Val),
              (Gly-Val-Gly), (Gly-Ala-Gly), (Gly-Leu-Gly)

Fig.1. The repetitive sequences in elastins.

Thus, elastin represents an ideal case of polypeptide chain whose conformational properties can be studied utilizing an approach
based on synthetic methods. Accordingly, a series of fragments
and analogs of elastin sequences was synthesized in our laborato
ry (as reported in Fig.2) in order to get further insight into
the molecular conformation of elastin.

i) poly(Val-Pro-Gly)

ii)poly(X-Gly-Gly); X=Ala, Val, Leu

iii) BOC-(Gly-Val-Gly-Gly-Leu)$_n$-OMe   n=1, 2, 3

Fig.2. The synthesized fragments and
analogs of elastin sequences.

In the followings we shall report only the general conclusions
derived from our rather extensive conformational studies on these synthetic compounds by discussing in some detail only few examples as the most representative ones.

RESULTS AND DISCUSSION

The NMR parameters of poly(Val-Gly-Gly) and poly(Leu-Gly-Gly),
(Table I), evidence the presence of a type II ß-turn having the
sequence -Val(Leu)-Gly- at the corners and two glycyl residues
connected by the 4→1 hydrogen bond.

Table I. Temperature coefficients, vicinal coupling constants, and relative torsional φ angles, of the NH protons of the polytripeptides in DMSO-D$_6$. The subscript i refers to the isomeric conformational peaks.

POLYPEPTIDE	RESIDUE	$\Delta\delta/\Delta T$ (ppb/K)	J(NHCH)(Hz)	φ^a
poly(Val-Gly-Gly)	Val^1	-6.5	7.3[b]	-156;-84
	Val^1_i	-4.1		
	Gly^2	-7.2	12.9	$62;54$
	Gly^2_i	-6.6		
	Gly^3	+0.7	9.9	141; 48
	Gly^3_i	-4.0		
poly(Leu-Gly-Gly)	Leu^1	-7.0	6.8	-159;84;68;52
	Leu^1_i	-5.0		
	Gly^2	-5.7	12.8	120;54
	Gly^2_i	-5.8		
	Gly^3	-2.0	10.0	141;48
	Gly^3_i	-4.7		

a) Values in italics are those compatible with standard type II ß-turn

b) The J's of the glycine residues are the sum of J_{AB} and J_{BX}.

The values of the temperature coefficients indicate large amounts of folded forms in the conformer population. Furthermore, the φ angles approximated from the vicinal coupling constants, comprise values quite compatible with standard type II ß-turns.

The following points should be appreciated:

i) slow conformational equilibria (on the NMR scale) are clearly discernible on inspection of the NH region. As a matter of fact, isomeric conformational peaks for all residues are present, suggesting a polypeptide chain with substantial mobility and flexibility.

ii) The amount of the isomeric peaks, corresponding to an ensemble of conformations is significantly higher for poly(Leu-Gly-Gly) than for poly(Val-Gly-Gly). More, the temperature coef ficient value of Gly[3] NH is lesser for poly(Val-Gly-Gly) than for poly(Leu-Gly-Gly). That the amount of the ß-turn strongly depends on the nature of the X residue is clearly confirmed by the circular dichroism (CD) spectra shown in Fig.3.

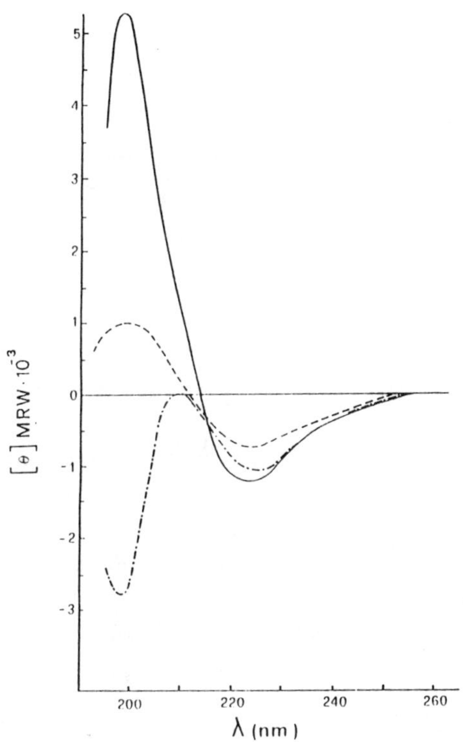

Fig.3. CD spectra in hexafluoroisopropanol
of poly(Val-Gly-Gly)(———), poly(Ala-Gly-Gly)
(---) and poly(Leu-Gly-Gly)(-·-·),c,0.2mg/ml.

Analogous studies carried out on the penta-, deca-, and pentadecapeptide reveal a more complex pattern. The conformation of these sequences, which are really present as dimeric(Raju and Anwar, 1987; Sandberg et al., 1985) and trimeric (Bressan et al., 1987; Yoon et al., 1984) species in elastins, seems to depend on the chain length. To this regard, the CD spectra, shown in Fig.4,are particularly impressive.

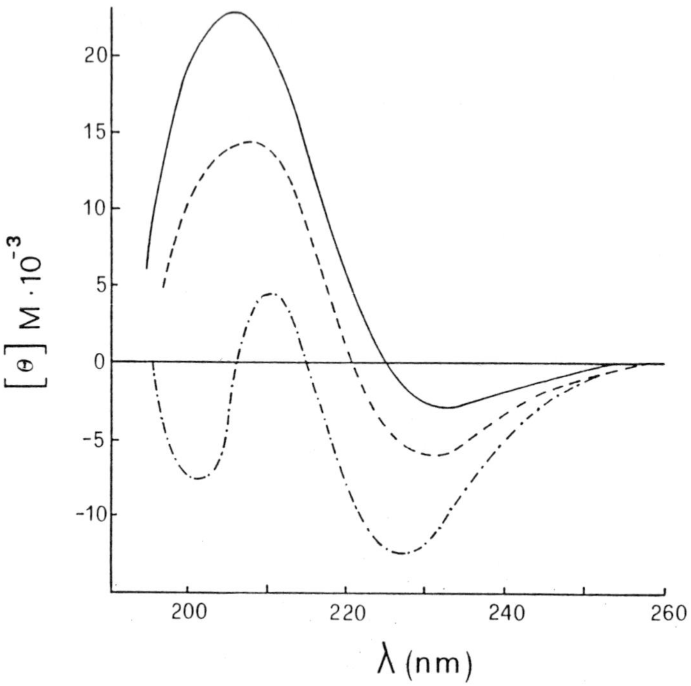

Fig.4. CD spectra in trifluoroethanol of
BOC-(Gly-Val-Gly-Leu)-OMe(——); BOC-(Gly-
-Val-Gly-Gly-Leu)$_2$-OMe(---); BOC-(Gly-Val-
-Gly-Gly-Leu)$_3$-OMe(-·-·) c,0.39 mM.

All curves are compatible with the presence, in the conformer po
pulation, of type IIß-turns. Quite interestingly, however, it can
be easily seen that the stability of the ß-turns decreases on goi
ng from the monomer to the dimer to the trimer. In the case of
the monomer, the NMR data are consistent with the presence of a
ß-turn having the sequence Val-Gly at the corners with the hydro
gen bond connecting Gly[1] and Gly[4] (Table II).

Table II. Temperature coefficients and vicinal coupling constants
of the NH protons of BOC-Val-Gly-Gly-Leu-OMe in DMSO-d$_6$

RESIDUE	$-\Delta\delta/\Delta T$(ppb/K)	J(NHCH) (Hz)
Gly^1	5.6	5.4
Val^2	3.7	8.3
Gly^3	4.6	5.9
Gly^4	4.5	5.9
Leu^5	4.8	7.3

However, the $\Delta\delta/\Delta T$ value for Gly^4 (-4.5 ppb/K) suggests a labi
le hydrogen bond and an equilibrium with non hydrogen-bonded con
formers. Furthermore, the data reported in Table II suggest also
other conformational features. Actually, the low temperature coef
ficient found for the NH of Val^2 indicates that also this group
partecipates to a possible weak hydrogen bond. The simplest inter
pretation is an additional C$_7$ cycle involving a hydrogen bond
between the NH of Val^2 and the CO of the BOC group (Fig.5).

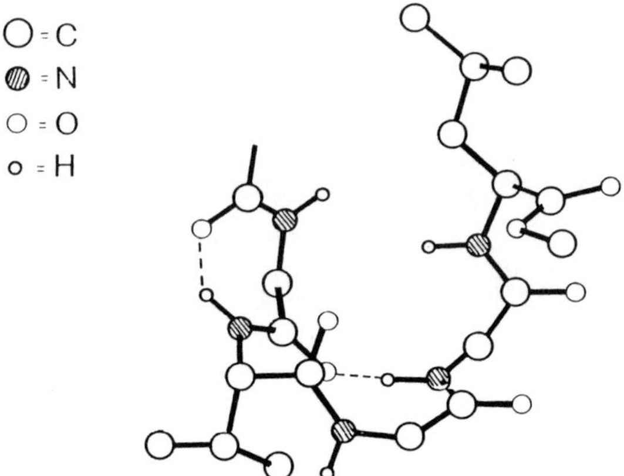

Fig.5. Proposed structure for BOC-Gly-Val-Gly-
-Gly-Leu-OMe showing the C$_{10}$ hydrogen-bonded
cycle(type II ß-turn) and the C$_7$ hydrogen-
bonded cycle. The (CH$_3$)$_3$-C-O moiety of t-
butyloxycarbonyl protecting group is not
shown.

Obviously, this feature should not be of relevance to the elastin conformation.

The rather low temperature coefficient found for Leu could be explained on the basis of an equilibrium involving two type II ß-bends:

$$\text{Gly}^1 \; \text{CO---HN Gly}^4 \qquad \text{Val}^2 \; \text{CO---HN Leu}^5$$

as if the ß-turns could "slide" along the polypeptide chain.

Finally, the fact that also other residues, such as Gly^3, exhibit moderately low temperature coefficients seem to indicate a complex, flexible pattern of labile hydrogen bonds which are exchanging their position along the sequence and giving rise to transient equilibria between folded and unfolded forms.

The φ torsional angles, approximated from J(NH CH) coupling constants, are not incompatible with the above suggestions, inasmuch as the multinumerical solutions of the Karplus-like equation were consistent throughout with both ß-turns having Val-Gly and Gly--Gly at the corners.

Quite interestingly, this peptide, analogously to α-elastin, shows an increased stability (Fig.6) of the ß-turn structure when its concentration is increased, therefore lowering the water activity.

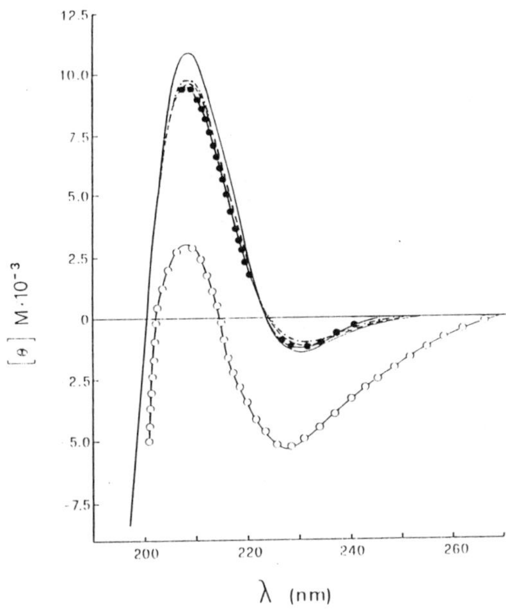

Fig.6. CD spectra of BOC-Gly-Val-Gly-Gly-Leu-
-OMe in water as a function of concentration:
(o-o-) 0.01 mg/ml; (——) 0.1 mg/ml; (---) 0.5
mg/ml; (-·-.) 0.8 mg/ml; (●-●-) 1.0 mg/ml.

On the other hand, the NMR spectra of the deca- and pentadecapeptide are indicative of increased amounts of unordered conformatio
nal states. They exhibited a very narrow chemical shift dispersi
on for the resonances of homologous residues, suggesting either
the lack of definite structure or a highly repetitive one. On the
other hand, two-dimensional double quantum and NOESY spectra for
the decapeptide, showed characteristic and regular NOE patterns
between the NH of a residue and the CH of the successive one, and
the lack of any other NOEs. This is only consistent with the pre-
dominance of rather extended structures. However, judging from the
NH temperature coefficients (Table III), one could suggest also

for the decapeptide low amounts of the following ß-turns in mutual equilibrium:

$$\text{Gly}^3 \text{ CO---HN Gly}^6 \qquad \text{Val}^7 \text{ CO---HN Leu}^{10}$$

Table III. Temperature coefficients of the NH protons of BOC-(Gly-Val-Gly-Gly-Leu)$_2$OMe and BOC-(Gly-Val-Gly-Gly-Leu)$_3$OMe in DMSO-d$_6$. The subsript i refers to the isomeric conformational peaks; the roman numbers indicate only the progressive indexing of the signals and do not refer to the sequence positions.

PEPTIDE	RESIDUE	$-\Delta\delta/\Delta T$(ppb/K)	PEPTIDE	RESIDUE	$-\Delta\delta/\Delta T$ (ppb/K)
DECAPEPTIDE	Gly^1	6.3	PENTADECAPEPTIDE	Gly^1	6.1
	Gly^1_i	1.5;1.6;1.5		Gly^1_i	1.9
	Val^2	4.8		Val^2	5.1
	Val^2_i	4.9		Gly^3	5.9
	Gly^3	4.8		Val^I	6.0
	Gly^4	6.0		Val^{II}	6.3
	Leu^5	5.4		Leu^I	5.1
	Gly^6	4.2		Leu^{II}	5.5
	Val^7	5.1		Leu^{III}	5.6
	Val^7_i	5.4		Gly^I	6.7
	Gly^8	5.2		Gly^{II}	6.3
	Gly^9	5.5		Gly^{III}	5.4
	Leu^{10}	4.6		Gly^{IV}	5.5
				Gly^V	6.2
				Gly^{VI}	5.5
				Gly^{VII}	5.5

On the contrary, for the pentadecapeptide the NMR results do not give any evidence for the presence of folded conformations.

Thus, the CD and NMR data, taken together, both indicate for the above sequences a decreased stability of the ß-turns as the poly peptide chain is lengthned.However, some apparent discrepancies are also evident. In fact, on NMR grounds, the presence of ß-turns in both deca- and pentadecapeptide is largely questionable, whereas their stability is not negligeable, although decreased, according to the CD data in trifluoroethanol. Obviously, the problem arises from the different solvents used in the two spectroscopies: DMSO is much more polar than the fluorinated alcohol and, therefore, the hydrogen bonds should be less stable in DMSO. CD spectra,carried out in the presence of the very polar solvent water, confirm this point showing that the stability of the ß-turns is strongly decreased for all considered peptides.

CONCLUSION

As far as the molecular structure of elastin is concerned, the most important aspects revealed by the reported studies are:

i)   the ß-turn conformation is the unique secondary conformation ubiquitously present in the repetitive regions of the protein. In fact this particular bend of the polypeptide chain is present in both X-Pro-Gly-Y and Gly-X-Gly-Gly containing regions of the protein;

ii) as a general trend, the most stable ß-turns contain the X-Gly sequence at the corners of the bend, the first and the fourth glycines being connected by a 4→1 hydrogen bond. This bond is of decreased stability in polar solvents such as DMSO and water, that is in those solvents where elastin displays its function (Gotte et al., 1968) and whose role of plastifiers has been previously suggested (Tamburro, 1981);

iii) the stability of the ß-turns depends on the nature of the X residue, on the hydration of the chain and, in the case of the sequence Gly-Val-Gly-Gly-Leu, is decreasing by increasing the length of the chain. As these are repetitive sequences in elastins, one should infer the presence of isolated ß-turns in these regions of the protein, at variance with the recurring ß-turns (the so called ß-spirals) proposed (Urry, 1988a) for the X-Pro-Gly-Y containing regions;

iiii)on the other hand, it should be appreciated that the structure of the protein is characterized by multiple equilibria between different conformational states, comprising dynamic ß-turns, other structures and also unordered states. In other words, many sequences, particularly those containing Gly-Gly bonds, are accessible to a rather large conformational space. As a matter of fact, it is not difficult to envisage a conformational equilibrium, comprising a multiplicity of intermediates, between folded conformations (such as ß-turns), changing their position along the sequence, and extended ones. This conformational transition should not involve gross energy barries, therefore suggesting an elasticity mechanism for elastin, at leastin part compatible with the classical one (Flory, 1969; Andrady and Mark, 1980).

Acknowledgments: the authors gratefully acknowledge the valuable aid of drs. M. A. Castiglione Morelli and A. Scopa.

REFERENCES

Abatangelo,G., Daga Gordini, D., and Tamburro, A. M. (1973) Int. J.Peptide Protein Res., 5 63-68.
Andrady, A. L., and Mark,J.E. (1980) Biopolymers 19 849-855.
Bressan, G.M., Argos,P., and Stanley, K.K. (1987) Biochemistry 26 1497-1503.
Gotte, L., Mammi, M., and Pezzin, G. (1968) In:Symposium on Fibrous Proteins, (W.G. Grewter, Ed.), Butterworths, Aus., pp. 236-245, and references quoted therein.
Guantieri, V., Jaques, A.M., Serafini-Fracassini, A., and Tamburro

A.M.   (1987) Biopolymers $\underline{26}$, 1957-1963.

Indik, Z., Yeh, H., Ornestein-Goldstein, N., Sheppard, P., Anderson, N., Rosenbloom, J.C., Peltonen, L., and Rosenbloom, J.
(1987) Proc. Natl. Acad. Sci. USA, $\underline{84}$, 5680-5684 .

Flory, P.J. (1969) Statistical Mechanics of Chain Molecules,
Wiley-Interscience, New York.

Partridge, S.M. (1977) In: Elastin and Elastic Tissue (L.B. Sandberg, W.R. Gray, and C. Franzblau, Eds),Adv. Exp. Med. Biol.,
$\underline{79}$, 715-721.

Raju, K., and Anwar, R. (1987) J. Biol. Chem., $\underline{262}$, 5755-5762.

Sandberg, L.B., Leslie, J.G., Leach, C. T., Alvarez, V.L., Torres,
A.R., and Smith, D.W. (1985) Pathol. Biol., $\underline{33}$, 266-274.

Tamburro, A.M., Guantieri, V., Daga-Gordini, D., and Abatangelo,
G. (1977) Biochim. Biophys. Acta, $\underline{492}$, 370-376.

Tamburro, A.M., Guantieri, V., Daga-Gordini, D., and Abatangelo,
G. (1978) J. Biol. Chem., $\underline{253}$, 2893-2894.

Tamburro, A.M. (1981) In: Connective Tissue Research: Chemistry,
Biology and Physiology (Z. Deyl and M. Adam, Eds.), A.R. Liss,
Inc., New York, pp. 45-62.

Urry, D.W. (1988a) J. Protein Chem., $\underline{7}$, 1-34, and references
quoted therein.

Urry, D.W. (1988b) J. Protein Chem., $\underline{7}$, 81-114, and references
quoted therein.

Yoon, K., May, M., Goldstein, N., Indik, Z.K., Oliver, L., Boyd,
C., and Rosenbloom, J. (1984) Biochem. Biophys. Res. Commun.,
$\underline{118}$, 261-269.

STRUCTURE AND DYNAMICS OF A ~19kD PHOSPHOPROTEIN, AMELOGENIN, FROM BOVINE
TOOTH ENAMEL

V. Renugopalakrishnan[a*], M. Prabhakaran[b], S. -G. Huang[c], H. C. Cheung[d],
E. Strawich[a], and M. J. Glimcher[a]

a. Laboratory for the Study of Skeletal Disorders and Rehabilitation, Dept. of
   Orthopaedic Surgery, Harvard Medical School and Children's Hospital,
   Enders-12, 300 Longwood Avenue, Boston, MA 02115, U.S.A.

b. Dept. of Medicinal Chemistry, College of Pharmacy, University of Illinois
   at Chicago, Chicago, IL 60680, U.S.A.

c. Dept. of Chemistry, Harvard University, Cambridge, MA 02138, U.S.A.

d. Dept. of Biochemistry, University of Alabama at Birmingham, Birmingham, AL
   35294, U.S.A.

SUMMARY:

Structure and dynamics of a hydrophobic protein from bovine tooth enamel
with a M_r ~19-20kD is derived from 2D NMR, Molecular mechanics-dynamics, and
time-resolved picosecond fluorescence studies. Amelogenin is postulated to
contain a hydrophobic core, β-spiral, encapsulated by a hydrophilic coat and
seems to act as a Ca^{++} pump in the early stages of enamel mineralization.

Amelogenin, an extracellular hydrophobic protein, with a M_r ~19-20 kD appears in the nascent stages of enamel mineralization (FIncham and Belcourt, 1985; Deutsch, 1989). The isolation, purification, and characterization of amelogenin from pre-molar calf teeth has been reported earlier from this laboratory, Strawich et al., 1985. Earlier, its amino acid sequence was reported by Edman degradation and gas-phase sequencing by Takagi et al., 1984. More recently, the primary structure has also been confirmed from its c-DNA sequence (Shimokawa et al., 1987). The amino acid composition of amelogenin is characterized by large proportion of Pro, Gln, Leu, His residues.

The primary structure of amelogenin is characterized by tandem repeats of Gln-Pro-X sequence interspersed with many Pro rich segments. A recent sequence search based on National Biomedical Research Foundation and EMBO data bases has revealed that the sequence bears little or no homology to any known protein sequence with one exception, RNA polymerases (Allison et al., 1985). While tandem repeats are known to occur in rhodopsin (Ovchinnikov et al., 1988), synexin and synaptophysin, two mammalian proteins implicated in calcium-dependent regulatory roles in the process of exocytosis (Creutz et al., 1980), gliadin, an abundant storage protein isolated from wheat kernel (Rafalski et al., 1984); hordeins, isolated from endosperm of barley, Hordeum vulgare (Forde et al., 1985), the tandem repeat occuring in amelogenin appears unique with only one exception RNA polymerases. RNA polymerase from Saccharomyces cerevisiae (Allison et al., 1985) contain multiple copies of a consensus heptapeptide, Tyr-Ser-Pro-Thr-Ser-Pro-Ser. The largest subunit, M_r~191 kD, of yeast RNA polymerase II, encoded by gene RPO21, has 27 tandem repeats of the consensus heptapeptide occurring to within 10 residues of its C-terminus. Corden et al., 1985 found 52 tandem repeats of the consensus hepta-peptide in the largest subunit of mouse RNA polymerase II. The discovery of RNA polymerase has for the first time revealed another protein in nature like amelogenin containing -X-Pro-X-X- tandem repeats.

From the beginning, structural studies of amelogenin experienced unusual difficulties arising from its poor insolubility in water and in aqueous buffers. CD spectra of amelogenin appeared to be devoid of readily recognizable structural elements (Renugopalakrishnan et al., 1986). It was not until Fourier transform infra-red spectroscopic studies (Renugopalakrishnan et al., 1986), the reason for the unusual spectroscopic patterns of amelogenin became somewhat clear. The spectral patterns were reminiscent of a protein containing low α-helical or β-sheet structures or a combination of both. Instead the spectral features were dominated by β-turns which seem to occur abundantly in the protein. The primary structure proposed initially by Takagi et al., 1984 indicated the rather high probability of β-turns caused by the occurrence of -X-Pro-X-X- tetrapeptides due to the strategic location of Pro residues, Fig. 1.

```
1                             10                                20
Met—Pro—Leu—Pro—Pro—His—Pro—Gly—His—Pro—Gly—Tyr—Ile—Asn—Phe—Ser—Tyr—Gln—Val—Leu—
                              30                                40
Thr—Pro—Leu—Lys—Trp—Tyr—Gln—Ser—Met—Ile—Arg—His—Pro—Tyr—Pro—Ser—Tyr—Gly—Tyr—Gln—
                              50                                60
Pro—Met—Gly—Gly—Trp—Leu—His—His—Gln—Ile—Ile—Pro—Val—Val—Ser—Gln—Gln—Thr—Pro—Gln—
                              70                                80
Asp—His—Ala—Glu—Gln—Pro—Pro—His—His—Ile—Pro—Met—Val—Pro—Ala—Gln—Gln—Pro—Val—Val—
                              90                                100
Pro—Gln—Gln—Pro—Met—Met—Pro—Val—Pro—Gly—Gln—His—Gln—Pro—Asn—Leu—Pro—Leu—Pro—Ala—
                              110                               120
Gln—Thr—Pro—Phe—Gln—Pro—Gln—Ser—Ile—Gln—Pro—Gln—Pro—His—Gln—Pro—Leu—Gln—Pro—His—
                              130                               140
Gln—Pro—Leu—Gln—Pro—Met—Gln—Pro—Leu—Gln—Pro—Leu—Gln—Pro—Leu—Gln—Pro—Leu—His—Pro—
                              150                               160
Pro—Gln—Pro—Ile—Val—Gln—Pro—Ile—Pro—Phe—Pro—Pro—Leu—Pro—Pro—Gln—Pro—Leu—Pro—Pro—
                              170
Met—Leu—Pro—Asn—Leu—Pro—Leu—Gln—Ala—Trp
```

Fig. 1 The primary structure of bovine amelogenin.

The spectral features of β-turns were less well understood in 1984 than today. From a knowledge of the primary structure of amelogenin, its CD and resolution enhanced FT-IR spectra became interpretable. A tentative model of amelogenin derived from Chou-Fasman predictions of the secondary structure consistent with experimental data is shown in Fig. 2.

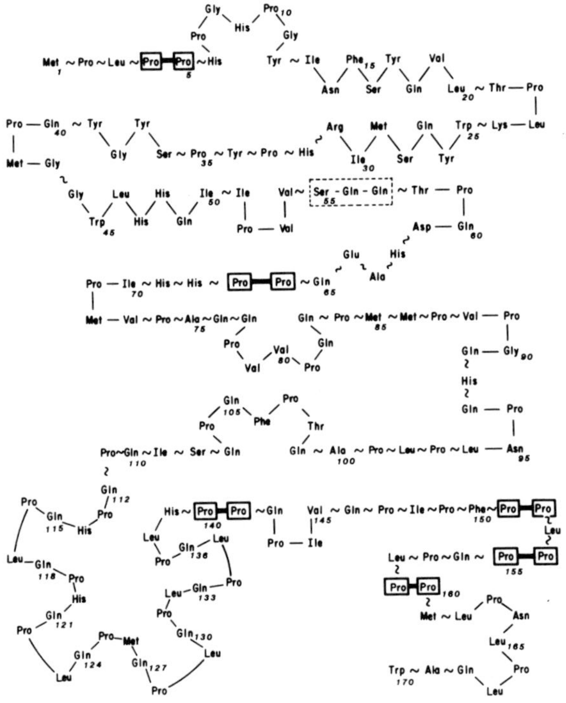

Fig. 2. The proposed secondary structure of bovine amelogenin.

The proposed structure was further corroborated by the appearance of a triplet amide I band with frequencies centered at 1664 cm $^{-1}$, 1674 cm $^{-1}$, and 1684 cm $^{-1}$, in the Raman spectrum of amelogenin in H_2O, characteristic of β-turns and β-sheet structures (Zheng et al., 1987). The conclusion derived from the Raman amide I triplet was further reaffirmed by complimentary amide III bands in the 1200-1300 cm $^{-1}$ domain.

At this stage we decided to investigate the secondary and tertiary structure of amelogenin by 2D NMR spectroscopy and molecular mechanics-dynamics approach (Renugopalakrishnan et al., 1988). While the 2D NMR studies were in progress, we also pursued the chemical synthesis of the tandem repeat, (-Gln-Pro-X-Gln-), between Gln_{112} through Leu_{138} by solid phase synthesis. The details of polypeptide synthesis and a report on its structure will appear elsewhere (Balasubramaniam et al., in preparation). More recently attempts are underway to attempt crystallization of the intact amelogenin and the tandem repeat. In addition, the expression of bovine cDNA clones in E. Coli to produce native amelogenin and its siblings are in progress collaboratively with Prof. Joel Rosenbloom, University of Pennsylvania, Philadelphia, PA.

2D NMR STUDIES OF AMELOGENIN

500 MHz ^{1}H NMR spectrum of amelogenin in 4:1-H_2O-D_2O was obtained and the details will be published elsewhere (Huang et al., in preparation). Proton assignments were systematically made by sequential assignment procedure (Wuthrich, 1986, Ch.7). The assignments are only partially complete as of August 1989. COSY spectrum of amelogenin in H_2O:D_2O-4:1 is presented in

Fig. 3.

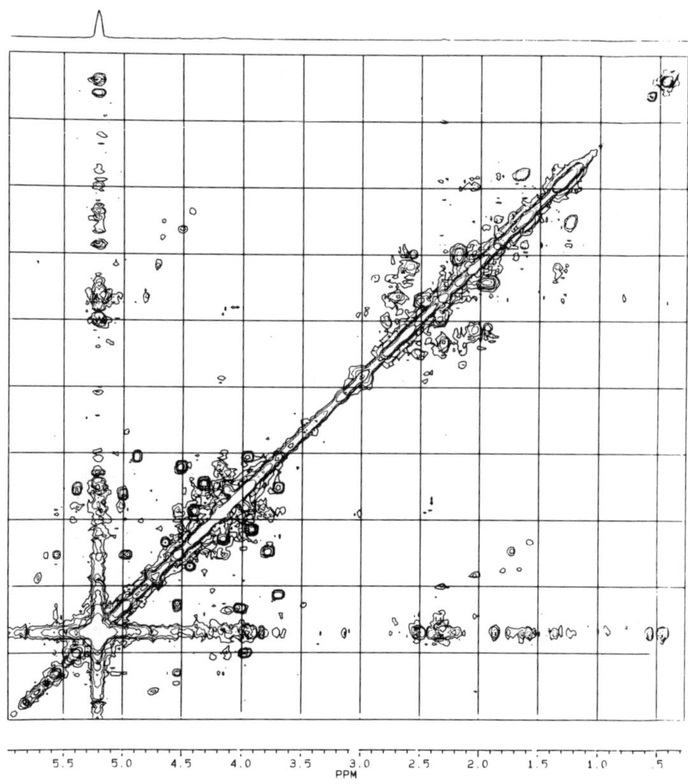

5.5 5.0 4.5 4.0 3.5 3.0 2.5 2.0 1.5 1.0 .5
PPM

Fig. 3. COSY spectrum of bovine amelogenin in $H_2O:D_2O$-4:1,
at approximately neutral pH.

The sequential assignment strategy was further supplemented by data derived

from ^{1}H NMR studies of numerous fragments which included Gln_{112}-Pro-His-Gln-

Pro-Leu-Gln-Pro-His-Gln-Pro-Leu-Gln-Pro-Met-Gln-Pro-Leu-Gln-Pro-Leu-Gln-Pro-

Leu-Gln-Pro-Leu-His_{139} which occurs at approximately the center of the

molecular and a C-terminal fragment (as deduced from amelogenin primary structure via cDNA sequence), -Leu-Pro-Asp-Leu-Pro-Leu-Glu-Ala-Trp-Pro-Ala-Thr-Asp-Lys-Thr-Lys-Arg-Glu-Glu-Val-Asp-Cys- obtained via genetic engineering methods.

Finally assignments were completed using 2D COSY experiments. 2D COSY spectrum of amelogenin in D_2O is shown in Fig. 3. The 2D NOESY experiments have provided a large number of inter proton connectivities which lucidly demonstrate the presence of numerous β-turns especially in the Gln_{112} through His_{139} segment. The β-spiral structure appears buried in the interior of the protein structure, shielded from the environment.

MOLECULAR MECHANICS-DYNAMICS STUDIES OF AMELOGENIN USING CRAY SUPER-COMPUTER

Molecular mechanics calculations were initially started on the tandem repeating segment, Gln_{112} -Pro-His-Gln-Pro-Leu-Gln-Pro-His-Gln-Pro-Leu-Gln-Pro-Met-$(Gln-Pro-Leu)_4$ by first resorting to model building based on stereochemical considerations. The occurrence of Pro residues in a generalized sequence, -X-Pro-X-X-, tetrapeptide led to the assumption that such a sequence has a high propensity for formation of β-turns. A type I β-turn was assumed to be formed. Families of helices were generated using the procedure of Sugeta and Miyawzawa, 1967 using reasonable starting $\{\phi, \psi\}$ values for the peptide backbone. The peptide groups were assumed to be planar with $\omega=180^0$. Amelogenin with a pI=7 (Deutsch, 1989) is postulated to play a crucial role in the nascent stages of enamel mineralization and yet its role is hard to explain due to very few Glu, Asp residues present in its primary structure and total absence of γ-carboxyglutamic(Gla) residues, the traditional chelators of Ca^{++} ions. Therefore it was postulated that peptide carbonyl groups of the polypeptide backbone by themselves sequester the Ca^{++} ions. Ca^{++} ion binding by neutral carbonyl group has been suggested previously by Urry, 1971 to explain similar mechanism thought to prevail in elastin and synthetic analogs and elastin peptides. In addition, previous studies have shown that β-turns act as loci for Ca^{++} ion binding (see Urry, 1978; 1983). With these two assumptions, family of helices generated by the methods of Sugeta and Miyazawa, 1967 were searched to find out whether the helical "pore" observed along the helical axis can physically accommodate the size of an ion like Ca^{++}. A limited number of helical structures, which we prefer to call β-spiral structure, were selected for refinement of the structure by molecular mechanics calculations. For our initial studies, program AMBER, developed in the laboratory of Prof. Peter A. Kollman, University of California, San Francisco, CA. (Singh and Kollman, 1984; Weiner et al., 1984) was used. The details of the calculations have been reported recently (Renugopalakrishnan et al., 1989).

Molecular dynamics calculations (van Gunsteren et al., 1983) were performed on the 27 residue polypeptide in the presence and absence of Ca^{++} ions. The positioning of Ca^{++} ions was accomplished by monitoring van der Waals and electrostatic interactions. Energy minimization along with low temperature dynamics resulted in an optimal environment for Ca^{++} ions along the helical axis.

The molecular dynamics calculations have now been extended to the entire protein molecule. A tentative model derived by a brute-force strategy indicates, that the dynamic β-spiral structure occurs in the middle of the protein with the helical "pore" buried in the interior but accessible from the outside, offers a novel mechanism for pumping of Ca^{++} ions into the mineralizing front in enamel.

TIME-RESOLVED PICOSECOND FLUORESCENCE SPECTROSCOPIC STUDIES

The emission decay of the Trp residues was determined in a photon-counting picosecond fluorometer (Hutnik and Zabo, 1989). A mode-locked Spectra Physics

15-W argon ion laser was used to synchronously pump a rhodamine dye laser at 82 MHz. The red output of the dye laser was cavity-pumped and frequency doubled through a potassium dihydrogen phosphate cyrstal and used as the excitation source. The repetition rate of the exciting pulses at 295 nm was 800 kHz. The decay measurements were carried out at 20°C with the sample dissolved in 50 mM Tris, pH 7.2, 1mM EGTA, in the absence or presence of 2mM Ca^{++}. The decay data was best fitted to a triexponential function with acceptable fitting statistics. The data suggest the presence of three populations of Trp residues.

Each population type is characterized by a decay time (τ_i) and a fractional intensity (f_i). At this time, it is difficult to decide whether the three decay times reflect 3 different secondary structures of amelogenin or 3 Trp residues. Addition of 2 mM Ca^{++} resulted in an increase of all decay times (τ_i), but because of the associated changes in fractional amplitudes the weighted average decay time actually decreased slightly. The fractional intensities (f_1 and f_2) of the two fastest components (τ_1 and τ_2) increased by 27 and 13% at the expense of a 7% decrease (f_3) of the slowest component (τ_3).

From the combined data a working model of amelogenin was derived which consists of a hydrophobic β-spiral encapsulated by a hydrophilic coat or exterior. The structural analysis is still in the process of further refinement but an overall view of the protein obtained so far explains its role as a Ca^{++} pump in the very early stages of enamel mineralization. In addition, amelogenin is also providing us insight into the mechanism of protein folding. Free energy calculations on the intact protein and its fragments are used as guide for site-specific mutagenesis.

Acknowledgement

This research was supported in part by National Institutes of Health grant AR34078.

REFERENCES

Allison, L.A., Moyle, M., Shales, M., and Ingles, C.J. (1985) Cell 42, 599-610.

Corden, J.L., Cadena, D.L., Ahearn, J.M., and Dahums, M.E. (1985) Proc. Natl. Acad. Sci. USA 82,7934-7938.

Creutz, C.E., Snyder, S.L., Husted, L.D., Beggerly, L.K., and Fox, J.W. (1988) Biochem. Biphys. Res. Commun. 152, 1298-1303.

Deutsch, D. (1989) The Anatomical Record 224, 189-210.

Fincham, A.G. and Belcourt, A.B. (1985) In: The Chemistry and Biology of Mineralized Tissues (Butler, W. T., Ed.), Ebsco Media Services, Inc., Birmingham, AL., USA, pp. 240-247.

Forde, B.G., Kreis, M., Williamson, M.S., Fry, R.P., Pywell, J., Shewry, P.R., Bunce, M., and Miflin, B. (1985) EMBO J. 4, 9-15.

Hutnik, C.M., and Zabo, A.G. (1989) Biochemistry 28, 3923-3934.

Ovchinnikov, Yu.A., Abdulaev, N.G., Zolotarev, A.S., Artamonov, I.D., Bespalov, I.A., Dergachev, A.E., and Tsuda, M. (1988) FEBS Lett. 232, 69-72.

Rafalski, J.A., Scheets, K., Metzler, M., Peterson, D.M., Hedgcoth, C., and Soll, D.G. (1984) EMBO J. 3, 1409-1415.

Renugopalakrishnan, V., Strawich, E.S., Horowitz, P.M., and Glimcher, M.J. (1986) Biochemistry 25, 4879-4887.

Renugopalakrishnan, V., Pattabiraman, N., Langridge, R., Strawich, E., Glimcher, M.J., Horowitz, P.M., Zheng, S., Tu, A.T., Huang, S. -G., Prabhakaran, M., Duzgunes, N., and Rapaka, R.s. (1988) In: Recent Progress in Chemistry and Biology of Centrally Acting Peptides (B.N. Dhawan and R.S. Rapaka, Eds.), Central Drug Research Institute, Lucknow, India, pp. 87-107.

Renugopalakrishnan, V., Pattabiraman, N., Prabhakaran, M., Strawich, E., and Glimcher, M.J. (1989) Biopolymers 26, 297-303.

Singh, U.C. and Kollman, P.A. (1984) J. Computational Chem. 5, 129-149.

Shimokawa, H., Ogata, Y., Sasaki, S., Sobel, M.E., MacQuillan, C.I., Termine, J.D., and Young, M.F. (1987) Adv. Dental Res. 20, 2-14.

Strawich, E., Poon, P.H., Renugopalakrishnan, V., and Glimcher, M.J. (1985) FEBS Lett. 184, 188-192.

Sugeta, H., and Miyazawa, T. (1967) Biopolymers 5, 673-679.

Takagi, T., Suzuki, M., Baba, T., Minegishi, K., and Sasaki, S. (1984) Biochem. Biophys. Res. Commun. 121, 592-597.

Urry, D.W. (1971) Proc. Natl. Acad. Sci. USA 68, 810-814.

Urry, D.W. (1978) Ann. NY Acad. Sci. 307, 3-27.

Urry, D.W. (1983) Ultrastructural Pathology 4, 227-251.

van Gunsteren, W.F., Berendsen, H.J.C., Hermans, J., Hol, W.G., and Posta, J.P.M., (1983) Proc. Natl. Acad. Sci. USA 80, 4315-4319.

Weiner, P.K., Singh, U.C., Kollman, P.A., Caldwell, J., and Cass, D.A. (1984) AMBER (Assisted Model Building with Energy Refinement), University of California San Francisco, Molecular Mechanics and Dynamics Program, University of California, San Francisco, CA, USA.

STUDIES OF PROTEIN HYDRATION BY DIRECT NMR OBSERVATION OF INDIVIDUAL

PROTEIN-BOUND WATER MOLECULES

Gottfried Otting and Kurt Wüthrich

Institut für Molekularbiologie und Biophysik, Eidgenössische Technische
Hochschule-Hönggerberg, CH-8093 Zürich, Switzerland

SUMMARY: A novel, improved scheme for two-dimensional NMR experiments with
proteins in H_2O solution was used to observe individual hydration water
molecules. In these experiments the location of the hydration waters is
determined from their nuclear Overhauser effects to distinct amino acid
residues. In basic pancreatic trypsin inhibitor (BPTI) four internal water
molecules which had been reported in three different crystal forms
(Deisenhofer & Steigeman, 1975; Wlodawer et al., 1984, 1987) were found to
be in the same locations also in the solution structure, with life times
with respect to exchange of the water protons in excess of 0.30 ns.
Additional NOE's with polypeptide protons located on the protein surface may
involve either hydration water molecules or hydroxyl protons of amino acid
side chains. Their total number is small compared to the number of NOE's
expected from the hydration water molecules identified in the crystal
structures of BPTI.

Protein hydration is a dominant factor in the stabilization of the

spatial molecular structure, and plays an essential role in the mechanisms

of protein-mediated reactions. Protein solvation has, on the other hand,

been evasive to experimental investigations, and detailed structural

descriptions are available only from diffraction studies with single

crystals (e.g., Deisenhofer & Steigemann, 1975; Wlodawer et al., 1984,1987;

Teeter, 1984). For a deeper understanding of structural and functional

properties of protein surfaces, a search for experiments capable of

providing more detailed information on protein hydration in solution is of

foremost interest.

Protein structure determination in aqueous solution depends on one's ability to record ^{1}H NMR spectra in H_2O (Wüthrich, 1986). To suppress the dominant solvent resonance, presaturation by selective water irradiation is commonly used (Wüthrich, 1986; Anil-Kumar et al., 1980). This procedure ensures uniform excitation over the entire spectrum, and has the additional advantage of conceptual simplicity. On the other hand it prevents direct observation of the solvent, or its interactions with the protein. For work with nucleic acids alternative solvent suppression schemes using semiselective excitation have long been introduced (Redfield & Gupta, 1971), and in two-dimensional ^{1}H NMR experiments using a semiselective observation pulse chemical exchange of labile protons in the polymer chain with the solvent was clearly manifested (Kearns et al., 1983; Hilbers et al., 1983; Boelens et al., 1985). Here we used a novel variant of this type of experiment to study the hydration of proteins aqueous solution (Otting & Wüthrich, 1989).

<u>RESULTS AND DISCUSSION</u>

The initial experiments resulting in the identification of individual molecules of hydration water bound to a protein in solution were performed with the basic pancreatic trypsin inhibitor (BPTI). The experimental evidence for the identification of protein-bound water molecules is shown in Fig. 1. It is based on the following. (i) The chemical shift of H_2O at 4°C is 5.01 ppm. In BPTI it does not coincide with any of the proton resonances originating from protons of the protein. As a consequence all the peaks in the cross sections of Fig. 1A must arise either from intermolecular transfer of magnetization between H_2O and the protein, or from intra-protein nuclear Overhauser effects (NOE) with labile protons of hydroxyl or carboxyl groups, which exchange sufficiently rapidly to appear at the chemical shift position of the solvent water. (ii) Transfer of magnetization from water protons to potentially labile protein protons could proceed via NOE's as well as via chemical exchange. Both mechanisms would lead to positive cross peaks in NOESY (two dimensional nuclear Overhauser enhancement spectroscopy in the laboratory frame), whereas ROESY (two-dimensional nuclear Overhauser enhancement spectroscopy in the rotating frame) cross peaks are positive for chemical exchange and negative for NOE-interactions (Bothner-By et al., 1984). The different sign for most of the peaks seen in Fig. 1 at 4°C then clearly indicates that these cross peaks manifest selective NOE's between protons at the chemical shift of the water and protons of distinct individual amino acid residues. Positive ROESY peaks due to chemical exchange are observed for the side chain amino proton resonances of lysine residues between 7 and 8 ppm, and for the resonance of the side chain OH of Tyr 35 at 10.07 ppm. (iii) The temperature dependence of the ROESY cross peaks was different for labile and nonlabile protein protons. At 68°C the exchange of the most rapidly exchanging amide protons (Wagner & Wüthrich, 1982a) is sufficiently fast to dominate the magnetization transfer with the water. In contrast, the ROESY cross peaks with nonlabile protons have negative sign throughout, characteristic of NOE's. Note also the line broadening and the shifts toward the water resonance position of the exchange peaks with Tyr 35 OH and the amino protons of lysine, which are observed when changing the temperature from 4°C to 68°C.

Overall, a series of NOESY and ROESY spectra recorded at 4°C, 36°C, 50°C and 68°C was used to assign the NOE's between protein resonances and the water signal (Otting & Wüthrich, 1989). The protein resonances involved in the cross peaks were identified by their chemical shifts (Wagner et al., 1987; Wagner & Wüthrich, 1982b). In Fig. 1 those resonances are indicated for which unambiguous assignments were obtained. In the crystal structures of BPTI (Deisenhofer & Steigemann, 1975; Wlodawer et al., 1984, 1987) the

<u>Fig. 1</u> Two-dimensional nuclear Overhauser enhancement (NOE) measurements in the laboratory (NOESY) and in the rotating frame (ROESY). The spectra were recorded at 600 MHz on a Bruker AM 600 spectrometer, using a 15 mM solution of BPTI in 90% H_2O / 10% D_2O, 100 mM NaCl, pH 3.5, with mixing times of 112 ms for NOESY and 87 ms for ROESY. Suppression of the water resonance was achieved by inserting a combination of selective pulses on H_2O, nonselective pulses and a short spin lock of 2 ms duration at the end of the mixing period (details in Otting & Wüthrich, 1989). Cross sections parallel to the ω_2-axis taken at the ω_1-frequency of the water resonance are shown at the two temperatures indicated. Cross peaks of polypeptide backbone amide protons with the water resonance are identified with the one-letter amino-acid symbol and the sequence number in the polypeptide chain. For other protons the proton type is also indicated. In the NOESY spectrum at 4°C the stars on the right indicate the location of two artifactual t_1 ridges from intense signals on the diagonal, and those near 9 ppm identify tails from intramolecular NOESY cross peaks of the amide protons of Thr 11 and Arg 39 with $C^\alpha H$ of Tyr 10 at 4.93 ppm, and $C^\alpha H$ of Cys 38 at 4.96 ppm, respectively. In the spectra at 68°C several spikes between 0.6 and 2.2 ppm represent t_1-noise from intense diagonal peaks.

144

A
4°C
NOESY
ROESY
OH Y35
R39
T11
C14
G36
E49
K41
T32
D50
C38
Y10
T54
CεH Y23
CαH P9
δCH2 P13
αCH2 G36
βCH2 Y10
βCH2 N44
γCH3 T32
βCH2 P9
10.0
9.0
8.0
7.0
4.0
3.0
2.0
1.0
0.0
ω2(ppm)

B
68°C
NOESY
ROESY
K46
OH Y35
R39
T11
C14
G36
C38
E49
K41
Y10
D50
D3
CεH Y23
CαH P9
δCH2 P13
αCH2 G36
βCH2 Y10
βCH2 N44
γCH3 T32
10.0
9.0
8.0
7.0
3.0
2.0
1.0
0.0
ω2(ppm)

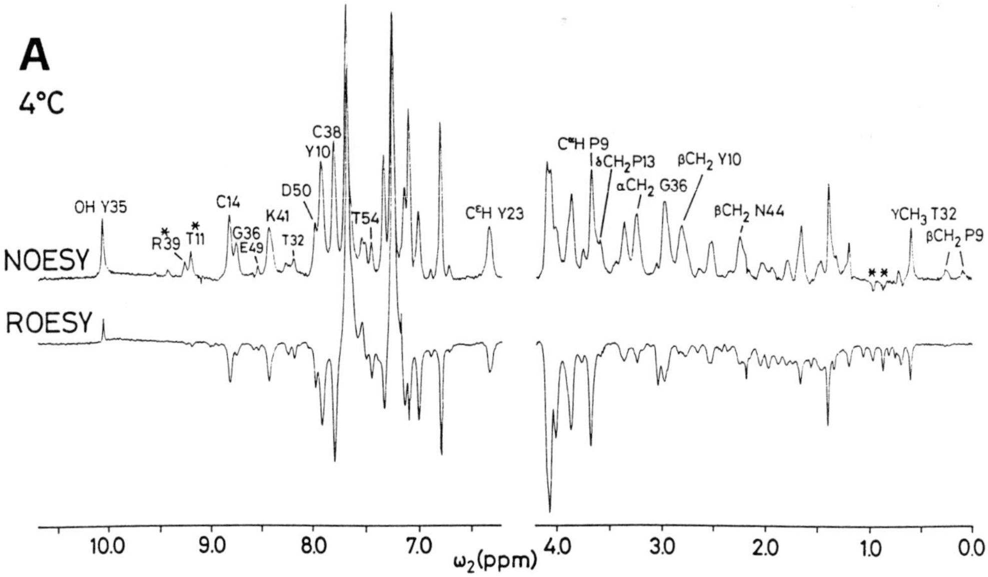

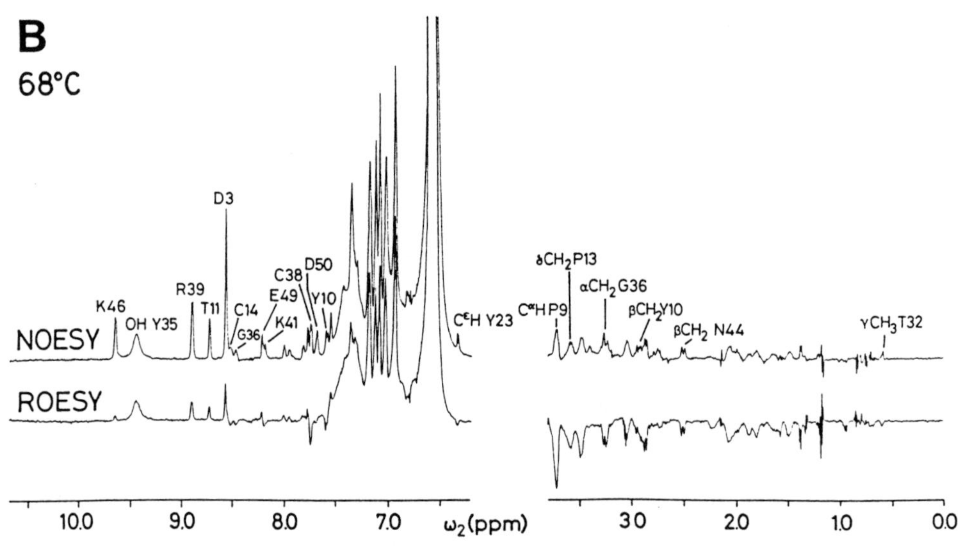

amide protons of Cys 14, Gly 36 and Cys 38, $C^{\delta}H_2$ of Pro 13 and $C^{\alpha}H_2$ of Gly 36 are all within 3.0 Å of the protons of the interior water molecule W122. Similarly, the amide protons of Tyr 10 and Lys 41, $C^{\alpha}H$ of Pro 9, $C^{\beta}H_2$ of Tyr 10 and $C^{\beta}H_2$ of Asn 44 are in close contact with one or two of the three interior water molecules W111, W112 and W113. The experiments of Fig. 1 thus provide direct evidence that these four interior water molecules are also present in the solution conformation of BPTI.

Additional NOE's were identified involving Thr 11, Thr 32, Thr 54, the aromatic ring of Tyr 23 , Glu 49 and Asp 50 (Fig. 1), which are all located near the protein surface. From the fact that positive rather than negative cross peaks manifest NOE's in NOESY (Fig. 1), we estimate that the water molecules which are involved in direct NOE's with protein protons must have life times with respect to exchange with the bulk water of $\gtrsim 3 \times 10^{-10}$ s. From earlier relaxation dispersion data on other proteins (Koenig et al., 1978) as well as from ^{18}O tracer techniques (Tüchsen et al., 1987), there is independent evidence for the presence of water molecules bound with a life time of this order of magnitude. Indirect evidence for negative nuclear Overhauser effects between water protons and protein protons has also been reported by Stoesz et al. (1978).

In the crystal structures of BPTI one finds approximately 60 hydration water molecules (Deisenhofer & Steigemann, 1975; Wlodawer et al., 1984, 1987). There are more than 200 distances between water protons and protein protons that are shorter than 3.0 Å, which should therefore in principle be detectable by NOE's. From Fig. 1 it is apparent that only a small subset of these, presumably comprising the most tightly bound H_2O molecules, gives rise to observable NOE's. Considering further that some of the cross peaks in Fig. 1 are probably due to intra-protein NOE's with hydroxyl or carboxyl protons in fast exchange with the bulk water, this demonstrates that in aqueous solution at most a small number of hydration sites can be characterized by life times that are comparable to those of the four interior water molecules with respect to exchange of the water protons with the bulk of the solvent.

The same experimental approach as described above for BPTI was also used to study several other small globular proteins. Overall, qualitatively similar results were obtained in all cases. With a scorpion toxin from <u>Androctonus australis Hector</u> several NOE's with a single internal water molecule could unambiguously be identified. For the <u>Antp</u> homeodomain of <u>Drosophila</u> we could observe only a few NOE's with the water line, which could either be attributed to surface hydroxyl groups, or which were too weak to present conclusive evidence for distinct hydration sites. For all proteins investigated so far the aromatic sidechains of Tyr give rise to intense cross peaks with the water signal, irrespective of whether they are located on the protein surface or in the interior of the protein. This finding supports that many of the NOE's observed at the water resonance position correspond probably to interactions with rapidly exchanging sidechain hydroxyl protons rather than with hydration water molecules. We are currently using three-dimensional NMR spectroscopy with uniformly ^{15}N-labeled proteins to assign further cross peaks between the water signal and protein resonances (Messerle et al., 1989).

In conclusion it should be added that our observations that at most very few surface hydration sites can be charaterized by NOE's is in agreement with molecular dynamics simulations, which calculated life times of the order of 10 ps for surface hydration sites (Ahlström et al., 1988 and references therein).

Acknowledgement: Financial support by the Schweizerischer Nationalfonds (project 31.25174.88) is gratefully acknowledged.

REFERENCES

Ahlström, P., Teleman, O., and Jönsson, B. (1988) J. Am. Chem. Soc. 110, 4198-4203.
Anil-Kumar, Wagner, G., Ernst, R.R., and Wüthrich, K. (1980) Biochem. Biophys. Res. Commun. 96, 1156-1163.
Boelens, R., Scheek, R.M., Dijkstra, K., and Kaptein, R. (1985) J. Magn. Reson. 62, 378-386.
Bothner-By, A.A., Stephens, R.L., and Lee, J. (1984) J. Am. Chem. Soc. 106, 811-813.
Deisenhofer, J., and Steigemann, W. (1975) Acta Crystallographica B31, 238-250.
Hilbers, C.W., Heerschap, A. Haasnoot, C.A.G., and Walters, J.A.L.I. (1983) J. Biomol. Struct. Dynamics 1, 183-207.
Kearns, D.R., Mirau, P.A., Assa-Munt, N., and Behling, R.W. (1983) In: Nucleic Acids: The Vectors of Life (B.Pullman and J.Jortner,Eds), Reidel Publishing Company, Dordrecht, pp. 113-125.
Koenig, S.H., Bryant, R.G., Hallenga, K., and Jacob, G.S. (1978) Biochemistry 17, 4348-4358.
Messerle, B., Wider, G., Otting, G., Weber, C., and Wüthrich, K. (1989) J. Magn. Reson., in press.
Otting, G., and Wüthrich, K. (1989) J. Am. Chem. Soc. 111, 1871-1875.
Redfield, A.G., and Gupta, R.K. (1971) J. Chem. Phys. 54, 1418-1419.
Stoesz, J.D., Redfield, A.G., and Malinowski, D. (1978) FEBS Lett. 91, 320-324.
Teeter, M.M. (1984) Proc. Natl. Acad. Sci USA 81, 6014-6018.

Tüchsen, E., Hayes, J.M., Ramaprasad, S., Copie, V., and Woodward, C. (1987) Biochemistry 26, 5163-5172.
Wagner, G., and Wüthrich, K. (1982a) J. Mol. Biol. 160, 343-361.
Wagner, G., and Wüthrich, K. (1982b) J. Mol. Biol. 155, 347-366.
Wagner, G., Braun, W., Havel, T.F., Schaumann, T., Go, N., and Wüthrich, K. (1987) J. Mol. Biol. 196, 611-639.
Wlodawer, A., Walter, S., Huber, R., and Sjölin, L. (1984) J. Mol. Biol. 180, 301-329.
Wlodawer, A., Nachman, J., Gilliland, G.L., Gallagher, W., and Woodward, C. (1987) J. Mol. Biol. 198, 469-480.
Wüthrich, K. (1986) NMR of Proteins and Nucleic Acids, Wiley, New York.

A ROLE FOR MOLECULAR MOTION AS A MECHANISM FOR THE NMR RELAXATION OF WATER PROTONS IN BIOLOGICAL SYSTEMS

C. Lin, D. W. Bearden, H. E. Rorschach and C. F. Hazlewood

Physics Department, Rice University, Houston TX 77251
Department of Physiology and Molecular Biophysics, Baylor College of Medicine, Houston, TX 77030

SUMMARY: The motion of water protons is changed due to their interactions with the macromolecules in the biological system. Our quasi-elastic neutron scattering studies show, through the increased correlation time in jump-diffusion model, that the diffusive motion of cellular water is reduced. This reduction in diffusive motion, however, cannot account for the observed reductions in the nuclear magnetic resonance (NMR) relaxation times and their frequency dispersion. In an attempt to explain the reduction in the NMR relaxation times (and the frequency dispersion), we have developed a physical model for the motion of proteins and their influence on water protons. The details of this model is discussed in this paper. We have been successful in interpreting some NMR frequency dispersion data of water protons in poly-amino acid and protein solutions with this theory. In addition, we have been successful in applying this theory to the dispersion of T_1 relaxation for water protons in cellular systems.

INTRODUCTION

Water exists in high percentage and plays an important role in all living cells. While water molecules still maintain a simple molecular structure in biological systems, their dynamic properties differ dramatically from the bulk state. The special and complicated behavior of water is part of the mystery of life.

The cause of these diversities is the presence of macromolecules and macromolecular structures, which disturb the homogeneity of the water phase. Through all kinds of interactions, the complex structure and motion of the macromolecular system change the structure and motion of the water surrounding them. These effects then spread out to the whole system due to the exchange and interaction within the water phase.

Among many experimental techniques, quasi-elastic neutron scattering (QNS)[1,2] and nuclear magnetic resonance (NMR)[3,4] are very effective for studing the motion of macromolecules and the dynamics of water. The combination of the results of QNS and NMR studies allow us to develop a physical picture for water-macromolecule interactions and the motion of water and macromolecules in the biological environment.

METHODS, MODELS AND RESULTS

1. QNS study and the jump-diffusion model

The QNS technique was used to study the dynamics of solvent water protons of deuteurized poly-ethylene oxide (dPolyox) by D. W. Bearden[5]. Thermal neutrons with energy E_1 and wavevector $\vec{k}_1$ incident on a sample may be scattered by a nucleus in the sample and leave the sample with final energy E_2 and $\vec{k}_2$. The differential cross section for this kind of scattering is given by[6]:

$$\frac{d^2\sigma}{d\Omega dE_2} = \frac{N}{4\pi} \frac{k_2}{k_1} \left\{ \sigma_c S(\vec{Q},\omega) + \sigma_i S_i(\vec{Q},\omega) \right\} \tag{1}$$

where $\hbar\omega = E_1 - E_2$ is the energy lost and $\vec{Q} = \vec{k}_1 - \vec{k}_2$ is the scattering vector. $S(\vec{Q},\omega)$, $S_i(\vec{Q},\omega)$ are the coherent and incoherent "scattering laws". In hydrogen containing system, the contribution is dominated by the incoherent scattering.

Many studies show that the hydrogen atom in water performs " jump diffusion"[6]. The hydrogen spends a residence time τ_0, on the average, at one site oscillating with mean-square amplitude $\langle u^2 \rangle$. The atom then diffuses to a new site within a time much shorter than τ_0, and starts oscillation at the new site. The incoherent scattering law $S_i(\vec{Q},\omega)$ for this type of motion is:

$$S_i^{jd}(\vec{Q},\omega) = (\pi\hbar)^{-1} e^{-Q^2\langle u^2\rangle/3} \frac{\Gamma/2}{\omega^2 + (\Gamma/2)^2} \tag{2}$$

with: $$\Gamma(Q) = \frac{2Q^2 D}{1 + Q^2 D \tau_0} \tag{3}$$

and: $D = \langle l^2 \rangle / 6\tau_0$. $\langle l^2 \rangle$ is the mean-square diffusion distance.

In our study, we measured quasi-elastic line with high resolution for 5%, 10%, 20% and 38% dPolyox in H_2O at different Q. Assuming there is a tight bound fraction of water associated with the polymer, we first fit the line shape with function:

$$I_0(Q,\omega) = A + B\omega + R(\omega)*\{ A_0(Q)\delta(\omega) + L(\Gamma_t) \} \qquad (4)$$

where $R(\omega)$ is a Gaussian resolution function and $A + B\omega$ represents the linear background. The bound fraction is represented by $A_0(Q)\delta(\omega)$, a δ-function with amplitude $A_0(Q)$. The free fraction is represented by $L(\Gamma_t)$, an Lorentzian with a FWHM Γ_t. But the fitting results for some of $A_0(Q)$s are negative. Others have large uncertainties. We also fit the data to a line shape function without the elastic scattering component:

$$I_0(Q,\omega) = A + B\omega + R(\omega)*L(\Gamma_t) \qquad (5)$$

Even though the χ_υ^2 values for both fitting are not significantly different. The errors for amplitudes and linewidth are much smaller with function (6). This suggests that the fraction of tight bound water is very small for dPolyox.

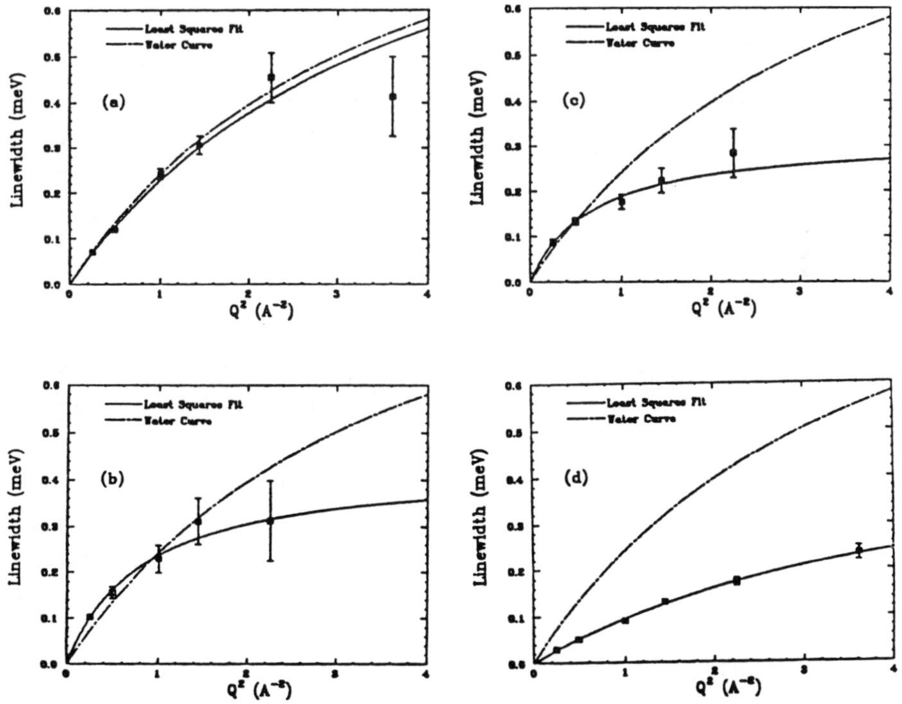

Fig. 1 Γ_t vs Q^2 for dPolyox in H_2O at (a) 5%, (b) 10%, (c) 20%, (d) 38% Concentrations.

By fitting the Q^2 dependence of Linewidth Γ_t obtained from fitting with function (6), we calculated the diffusion parameters of water in dPolyox solutions with different concentration (Table 1).

Table 1

Sample	χ_v^2	$D \times 10^5$ (cm^2/sec)	$\tau_0 \times 10^{12}$ (sec)	$\langle l^2 \rangle$ ($\overset{o}{A}{}^2$)
5% dPolyox in H_2O	1.66	2.21±0.07	1.22±0.07	1.62±0.13
10% dPolyox in H_2O	0.29	4.04±0.21	3.07±0.16	7.44±0.55
20% dPolyox in H_2O	0.40	3.59±0.30	4.18±0.35	9.00±0.11
38% dPolyox in H_2O	0.64	0.89±0.03	2.57±0.10	1.37±0.16

Comparing with the diffusion parameters for pure water[7], $D \times 10^5$ = 2.38 cm^2/sec, $\tau_0 \times 10^{12}$ = 1.2 sec and $\langle l^2 \rangle$ = 1.7 $\overset{o}{A}{}^2$, only the 38% dPolyox sample shows the decrease in diffusion coefficient, but the residence time increased for all the samples. The reason for obtaining large diffusion coefficients in 10% and 20% dPolyox samples is that the diffusion coefficient is determined by the initial slope of Γ_t vs Q^2. The linewidth at low Q has large uncertainty due to the poor statistics. Also the rotational motion of water molecules which is significant at low Q was not taken into account. The decrease of diffusion coefficient in dPolyox solutions is confirmed by NMR studies[8]. On the other hand, the residence times are determined by the linewidth at the high Q limit. Fig. 1 shows that the residence times in 10%, 20% and 38% solutions are much longer than for pure water.

The QNS study shows that the water in the polymer solutions still appears to be homogeneous. There is a very small amount of water tightly bound with the polymer that does not give a significant elastic scattering contribution. But, within the range of the interaction from the polymer, there is a shell of water (the hydration shell) which is different from the normal water by the increased hydrogen residence time. A "dual site" jump-diffusion model may provide a better picture for the motion of hydrogen in polymer solutions.

In the "dual site" jump-diffusion model, a hydrogen atom spends its time in both the hydration shell and the normal phase away from the polymer. The increase of the average residence time depends on the ratio of the population and on the ratio of residence times between these two types. Because the exchange of hydrogen between these two types of sites is very fast (on the order of the residence time), the water in polymer solution appears to be a single phase.

While the details of this model need further study, it gives two Lorentzian components associated with the fraction in the hydration shell and the free fraction. It explains why the initial slope of Γ_t vs Q^2 curve for polymer solutions differs little from the slope for pure water but levels off at the high Q limit.

2. NMRD studies and protein-dynamics model

NMR is another useful technique to study the dynamics of water. Many NMR studies have been completed on the properties of water in biological systems, especially relaxation time studies, because the difference in water relaxation time is directly related to the contrast in NMR imaging. Unfortunately, due to our poor knowledge of the dynamics of water and the interaction between water and macromolecules in biological system at present time, the basic mechanism for water relaxation remains elusive.

Relaxation studies made at various frequencies have revealed an interesting property: the frequency dependence of relaxation time has a close connection with the macromolecular structure in the system. A study in Dr. S. Koenig's laboratory on NMR dispersion in protein solutions showed that the dispersion increases as the molecular weight of the protein is increased[3] There are several suggestion for the mechanism of the frequency dispersion of the relaxation time for macromolecular solutions[9]. One group proposes a kind of slow motion, such as that due to rotational motion of "bound" water, which yields a single long correlation time. This type of model predicts a Lorentzian dispersion function which has ω^{-2} frequency dependence at the high-frequency limit, while the experimental data shows a $\omega^{-1/2}$ frequency dependence[3,9].

Based on the $\omega^{-1/2}$ frequency dependence, which requires relaxation modes with a spectrum of relaxation times and what we have learned from the quasi-elastic neutron scattering studies, we proposed the protein-dynamics model[10] to explain the water relaxation in macromolecular system.

In macromolecular solutions, the chain segments of macromolecules, which may have some water tightly bound, perform the highly-damped vibrational motion under the thermal excitation governed by the wave equation:

$$\sigma\frac{\delta^2 y}{\delta t^2} + \frac{1}{\mu}\frac{\delta y}{\delta t} - f\frac{\delta^2 y}{\delta z^2} = 0 \qquad (6)$$

where σ = segment mass/length, μ = segment mobility/length and f is chain tension. For a vibrational mode with wave number q, the correlation time is:

$$\tau_c = \frac{1}{f\mu q^2} \qquad (7)$$

The fluctuating magnetic field generated by these vibrations has the same spectrum of correlation times. Due to the electro-magnetic interaction between the fixed hydrogens along the chain and the hydrogens in the hydration shell, the hydrogen that diffuses into this surrounding shell changes its spin first. With the fast exchange between the hydration shell and the normal phase, the whole water phase relaxes with the same rate.

The relaxation rate for hydrogens moving with the vibrational chain can be calculated by summing over the contributions from all the vibration modes.

$$R_{1p} = \frac{2}{3}\gamma^2 \bar{h}^2 \int_{\tau_{cmin}}^{\tau_{cmax}} \frac{g(\tau_c)\tau_c d\tau_c}{1 + \omega^2\tau_c^2} \qquad (8)$$

where $g(\tau_c)$ is the mode density. For the frequency range $\tau_{cmin}^{-1} \gg \omega \gg \tau_{cmax}^{-1}$, we have:

$$R_{1p} = (2\pi/3\sqrt{2})\gamma^2 C\bar{h}^2\omega^{-1/2} \qquad (9)$$

Considering the cross relaxation and exchange, the frequency dispersion for protons in macromolecular solution has a form of $R = A\omega^{-1/2} + B$, as observed in many studies.

This theory was applied to mouse muscle data and on Artemia Cysts, with detailed discussion about the exchange process, by T. F. Egan[11]. For the further testing of this model, we also studied the R_1 dispersion for various poly-amino acids (poly-1-lysine [PL], poly-1-glutamic acid [PGA] and poly-proline [PP]) with different molecular weights (3.7k and 78K for PL, 51k and 13.6k for PGA and 19k for PP), and in the solutions with different pHs (Hi = 11.0 and Lo = 2.6). The data shown in the following figures is only for 51k poly-l-glutamic acidand poly-proline.

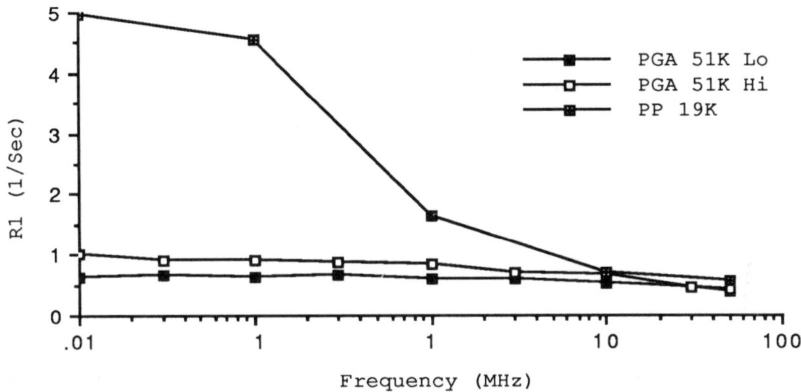

Fig. 2 1/T_1 dispersion for poly-glutamic acid And poly-proline.

The dispersions for poly-glutamic acid and poly-lysine are quite modest. The dispersion for poly-proline is much stronger. In fact, after the scaling, the dispersion for poly-proline is almost like the dispersion obtained by Koenig for 460k homocyanin[3], while the molecular weight of Poly-Proline is only 19k.

Because the side-chain structure doesn't allow the poly-proline to fold[12], it maintains an extended state in the solution. The effective chain-length is much longer than chain-length for globular protein with large molecular weight. The abundant low frequency vibrational modes lead to a fast relaxation rate at low frequency.

For poly-glutamic acid, the Coulomb repulsion between the charged groups[12] on the polymer causes a high tension f. The result is a higher low frequency cut off for the vibration modes, resulting in a small dispersion.

DISCUSSION

In the context of the protein-dynamics model, the highly-damped thermal vibrational motion produces a fluctuating magnet field on protons tightly bound to the macromolecule chain segment. When a water proton diffuses into a site in the hydration shell, it changes the orientation of its magnetic moment due to the interactions with the atoms associated with the vibrating chain. The fast exchange between this hydration shell and the bulk phase then cause all the water to relax. This is the basic mechanism for the relaxation of water proton in macromolecular system.

The vibrational modes of macromolecule depend on its structure and conformation as well as the other physical parameters such as chain tension. All this may change due to the interaction with the solvent molecules. While this idea provides a mechanism for relaxation, more study needs to be done before we can fully understand the relaxation mechanisms in complicated biological systems.

<u>REFERENCES</u>

1. E. C. Trantham, H. E. Rorschach, J. S. Clegg, C. F. Hazlewood, R. M. Nicklow and N. Wakabayashi, Diffusive Properties of Water in Artemia as Determined form Quasi- Elastic Neutron Scattering Spectra, <u>Biophys. J.</u> 45:927 (1984)
2. H. E. Rorschach, D. W. Bearden, C. F. Hazlewood, D. B. Heidorn and R. M. Nicklow, Quasi-Elastic Scattering Studies of Water Diffusion, <u>Scanning Microscopy</u> 1:2043 (1987)
3. K. Hallenga and S. H. Koenig, Protein Rotational Relaxation as Studied by Solvent ^{1}H and ^{2}H Magnetic Relaxation, <u>BioChem</u> Vol 15, No. 19, 4255 (1976)
4. P. A. Bottomly, T. H. Foster, R. E. Argersinger and L. M. Pfeifer, A Review of ^{1}H Nuclear Magnetic Resonance Relaxation in Pathology: Are T_1 and T_2 Diagnostic?, <u>Med. Phys.</u> 14:1 (1987)
5. D. W. Bearden, A Quasi-Elastic Neutron Scattering Study of Hydrogen Dynamics in Aqueous Polymer Solution, Ph. D. thesis, Rice University (1986)
6. T. Springer, Quasi-Elastic Neutron Scattering for the Investigation of Diffusive Motion in Solid and Liquid, 64, Springer-Verlag, Berlin (1972)
7. E. C. Trantham, Quasi-Elastic Neutron Scattering Study of Water in Agarous Gels, Ph. D. thesis, Rice University (1980)
8. D. W. Bearden, The Measurement of The Diffusion Coefficient of Water in Poly(Ethylene Oxide)/Water Solution Using a Nuclear Magnetic Resonance Pulsed Field Gradient Technique, M. A. Thesis, Rice University (1983)
9. M. F. Brown, J. F. Ellena, C. Trindle and G. D. Williams, Frequency Dependence of Spin-Lattice Relaxation Times of Lipid Bilayers, <u>J. Chem. Phys.</u> 84(1) (1986)
10. H. E. Rorschach and C. F. Hazlewood, Protein Dynamics and the NMR Relaxation Time T_1 of Water in Biological System, <u>J. Mag. Res.</u> 70:79 (1986)
11. T. F. Egan, Molecular Basis of Contrast in MRI, Cell Function and disease, Canedo *et al*, Plenum Press (1989)
12. A. L. Lehninger, Biochemistry, 2nd Edition, Worth Publisher Inc., (1975)

EFFECT OF COLLAGEN CROSSLINKING ON COLLAGEN-WATER INTERACTION

Madeleine BONNET, J. KOPP[†], J.P. RENOU

Station de Recherches sur la Viande, INRA-Theix, 63122 Ceyrat, France

SUMMARY:

The effects of collagen hydration and crosslinking state on protein-water interaction were investigated by Differential Scanning Calorimetry (DSC) and ^{1}H Nuclear Magnetic Resonance (NMR).

INTRODUCTION

Considerable works were performed to elucidate the state of water in biological systems (Mrevlishvili & Privalov, 1969; Susi et al., 1978; Lynch & Webster, 1975; Chein, 1975; Pineri et al, 1978; Hazlewood et al, 1988)

Collagen is the major structural protein of connective tissue. Toughness of meat is, at first, proportional to its intramuscular collagen content (Kopp & Bonnet, 1982). An equally important factor is the amount and the chemical nature of covalent crosslinks between collagen molecules (Kopp & Bonnet, 1986). With aging, these crosslinks become increasingly thermostable so that toughness of muscle increases (Kopp, 1971; Ledward, 1984; Bailey, 1985).

We studied collagen extracted from bovine muscles. At first, control tissue was purified with trypsin; less cross-linked one was this control tissue depolymerised with penicillamin; and the more crosslinked collagen was the same control reduced by borohydride (Bailey et al, 1970). In a second experiment we used three collagens extracted from veal, steer and cow muscles.

All connective tissues were hydrated under six well defined Water Activity (Aw) conditions from 0.440 to 0.972.

The effects of collagen hydration and crosslinking state were investigated by Differential Scanning Calorimetry (DSC) (Kopp et al, 1989) and [1]H Nuclear Magnetic Resonance (NMR) (Renou et al, 1983).

I - DIFFERENTIAL SCANNING CALORIMETRY (DSC) allows to determine:

- Amount of unfreezable water in the sample,

- Enthalpy change (ΔH) of collagen denaturation,

- Temperatures of transition of collagen to gelatin.

- <u>METHOD</u> (fig.1)

• First we recorded ice melting endotherm, from 223 to 293K. From the area of this peak we calculated corresponding amount of freezable water, and therefore amount of unfreezable water if Dry Matter (DM) of the sample had been measured.

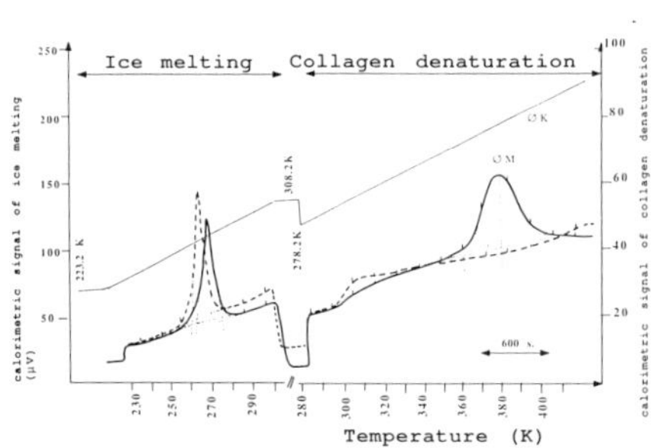

fig 1: thermogram of hydrated collagen (control tissue)

• Then we recorded endotherm of collagen denaturation, from 278 to 373K or more, according to amount of water contained in the sample. From the area of this peak we calculated denaturation enthalpy change; we read also maximum temperature of transition from collagen to gelatin.

- <u>RESULTS</u>:

For the three tissues we showed that **when moisture level** (Wt) **increased**:

• maximum temperature of denaturation decreased according to a decreasing exponential (fig 2): $(\emptyset_M)$ °C $= P_{1T} * (exp^{(-Wt * P_{2T})}) + P_{3T}$

• denaturation enthalpy change increased according to increasing exponential (fig 3): ΔH (J/g) $= P_{1H} (1- exp^{(-Wt * P_{2H})}) + P_{3H}$

• amount of unfreezable water increased according to an increasing exponential (fig 4): Wu (g/g DM) $= P_{1W}(1- exp^{(-Wt * P_{2W})}) + P_{3W}$

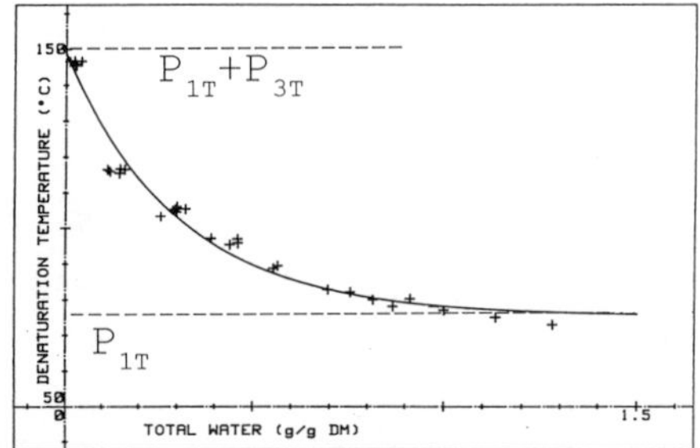

$$(\emptyset_M) °C = P_{1T} * (exp^{(-Wt * P2T)})+ P_{3T}$$

fig 2: Decrease in maximum denaturation température (ØM) with increasing moisture level for control tissue.

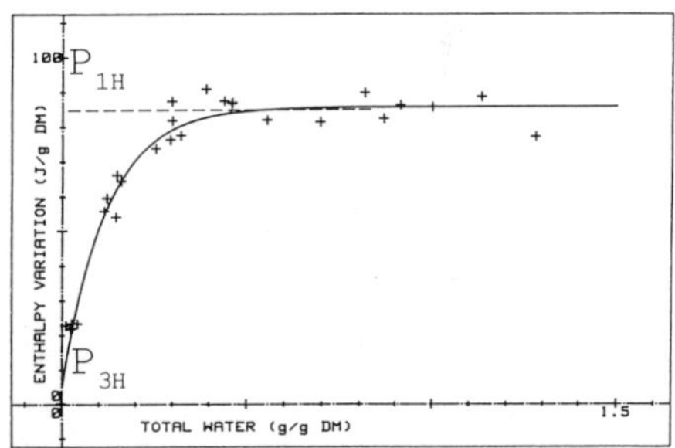

$$\Delta H = P_{1H} (1- exp^{(-Wt * P2H)}) + P_{3H}$$

fig 3: Increase in denaturation enthalpy with increasing moisture level.

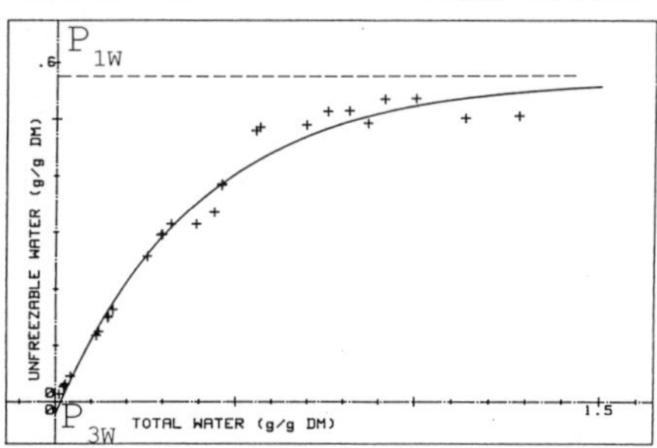

$$Wu^{.} = P_{1W}(1- exp^{(-Wt * P2w)}) + P_{3W}$$

fig 4: Increase in unfreezable water with increasing moisture level.

From these experiments we showed, that the less connective tissues were cross-linked, the more they could bind water:

More unfreezable water can be found in the less polymerized collagen than in other tissues (0.629 g water / g DM), therefore for the more crosslinked tissues, unfreezable water limit is only 0.552 (fig 5).

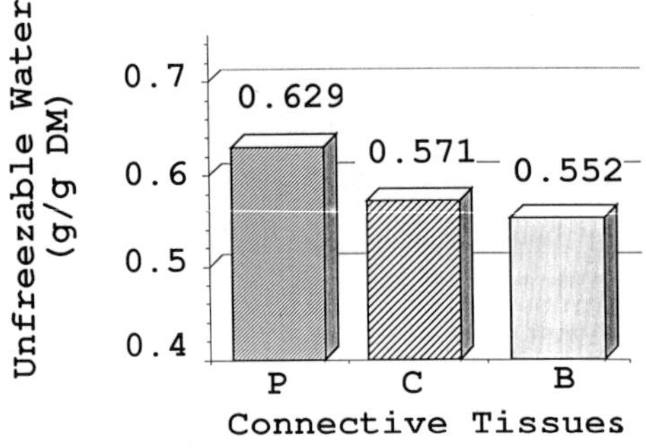

fig 5:Calculated maximum limit of the exponential increasing of unfreezable water in function of total water contained in samples, (P) penicillamin depolymerized, (C) control and (B) borohydride reduced tissues.

- From these results it appeared that water of collagen hydration could be classified into four states, the calculated limit of which varied according to the degree of crosslinking (fig 6):

• Firstly, the whole water is unfreezable, below 0.30 g/g DM, and very litle influenced by reticulation.

• Then, freezable water appears, but the whole water is associated with the protein so that denaturation enthalpy increases with total water content, the more collagen is cross-linked the smaller is this amount of water: from 0.30 to 0.60 for penicillamin treated tissue and from 0.28 to 0.44 for the most cross-linked one.

• Next, the water is probably less associated with the protein since denaturation enthalpy no longer increases, but denaturation maximum temperature still decreases. This amount of water is greater for the most cross-linked tissue, from 0.44 to 1.70, than for the less one (from 0.60 to 1.60)

• Finally, water which no longer induces modifications of denaturation characteristics but for which unfreezable water is still increasing. This state is the same for control and less polymerized tissues, we did not find it for the most cross-linked one.

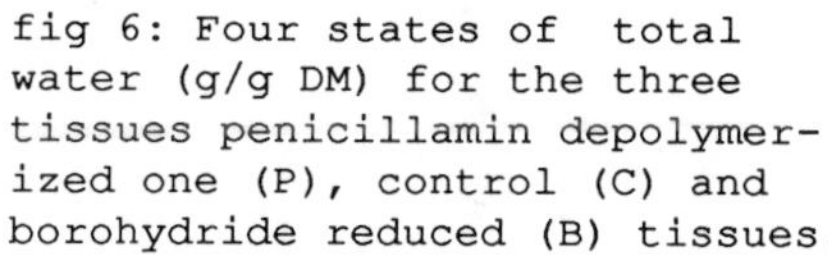

fig 6: Four states of total water (g/g DM) for the three tissues penicillamin depolymerized one (P), control (C) and borohydride reduced (B) tissues.

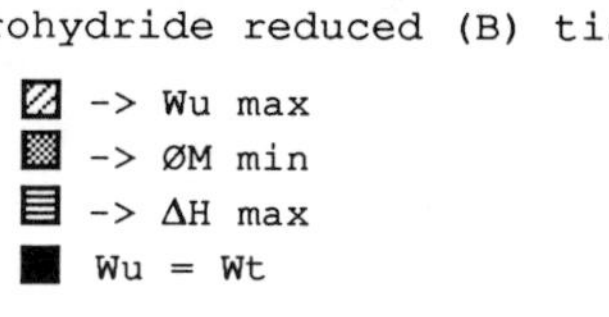

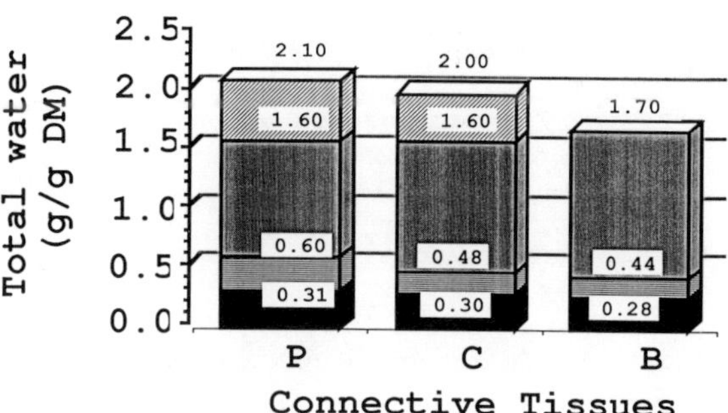

II - ^{1}H NUCLEAR MAGNETIC RESONANCE (NMR):

- METHOD:

Water proton NMR relaxation rates were monitored with a 400 MHz spectrometer, from 253 to 298K, by step of 5°; this for the three connective tissues, with the six hydration states.

We used the Goldman-Shen pulse sequence (Goldman & Shen, 1966), this method yields the cross-relaxation rate R^+ between water and proton collagen (fig 7).

After the time t0 collagen magnetization goes to zero. The second pulse in opposite phase with the first one brings only water magnetization to the steady magnetic field. The collagen and water magnetizations vary with time t under the effect of cross-relaxation.

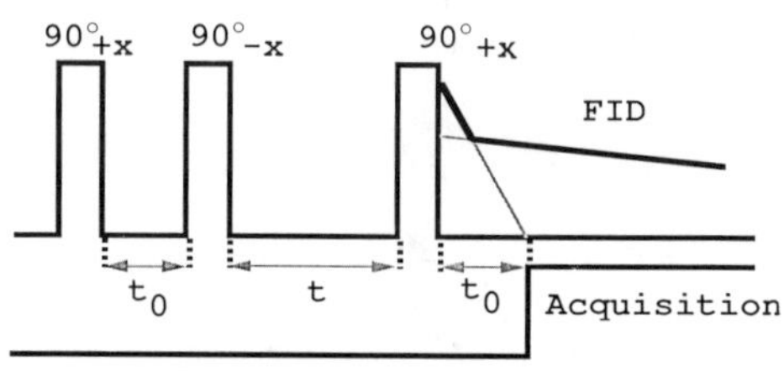

fig 7:Goldman-Shen pulse sequence

The third pulse is used to measure the free induction decay which is acquired after the time t0. After Fourier transformation, the intensity of the signal is recorded. The intensity of this signal varies as a function of t, in the same way as the difference of two exponentials (fig 8):

162

$$h(t) = A * \exp^{(-R_+ * t)} + B * \exp^{(-R_- * t)}$$

The decrease in a constant rate R^+ is due to the cross relaxation between both systems.

fig 8: Variation of ^1H RMN intensity signal as a function of time t

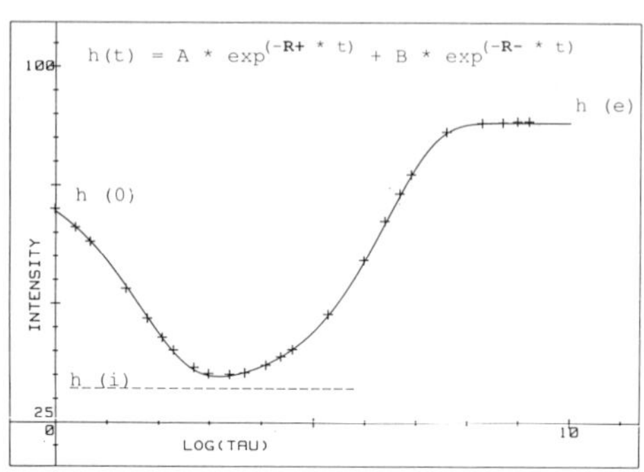

- <u>RESULTS:</u>

- Cross relaxation rates R^+ were water content and temperature dependent:

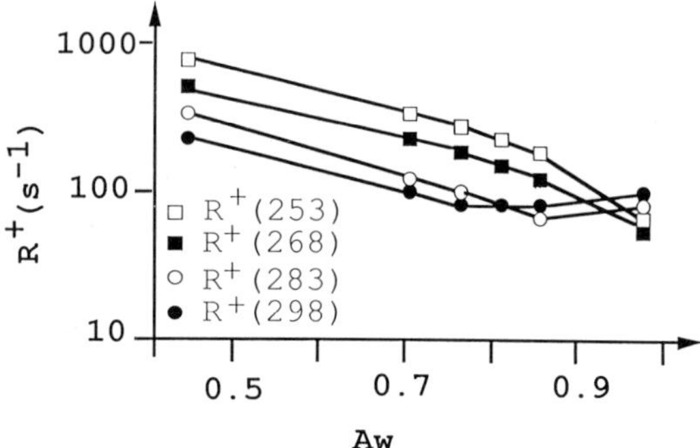

fig 9: R+ (s-1) as function of Aw for different temperatures. (cow collagen at four temperatures)

1) For low hydrations R+ decreased and then increased when Aw is greater than 0.85 (total water > 0.30 g/g DM). But this limit varied with temperature and cross-links: R+ increased from 273 K for cow and steer tissues and only from 278 K for veal collagen.

2) R+ showed a minimum for Aw = 0.972 (i.e. for total water g/g DM > 0.5), at 258 K. For other Aw this minimum exists probably, but then for temperatures over 298 K.

fig 10: R+ (s-1) as function of inverse temperature (1000/T K), for various Aw.

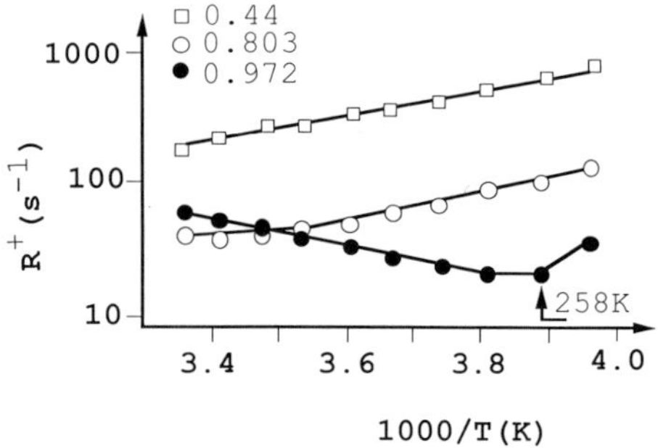

- We compared biochemical test of collagen reticulation (i.e. % of solubility of collagen when it was heated at 90°C during six hours) with cross-relaxation rates R+ at various temperatures and for different water contents.

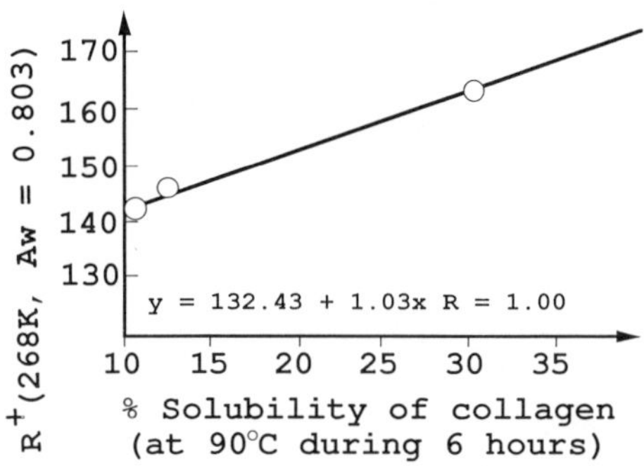

fig 11: linear relation between R+ and collagen solubility for Aw = 0.803, at 268 K.

For Aw and temperatures studied, significant linear relation between R+ and collagen solubility is the best for Aw = 0.803, at 268 K, In these conditions we have the relation (fig 11):

R+ = 132.43 + 1.03 (% solubility)

That is to say that, in theses conditions, the less collagen is cross-linked the greatest is the rate of cross-relaxation.

This should be verified for a larger number of polymerized collagens.

CONCLUSION

^{1}H NMR and DSC are complementary methods to characterize hydration and cross-linking states of native collagen.

DSC showed that the less connective tissues were cross-linked the more they could bind water. It showed also that the water of collagen hydration could be classified into four states whose limits varied according to the degree of crosslinking.

NMR had the advantage to allow study of water behaviour before any water freezes, i.e. below 0.30 g water / g Dry Matter. The different motions were temperature and water dependent.

In order to test the ability of both NMR and DSC parameters to characterize collagen crosslinking state, Discriminant Factorial Analysis was performed.

Cross-linking states were fully discriminated by the measurement of unfreezable water by DSC and of NMR cross relaxation rates

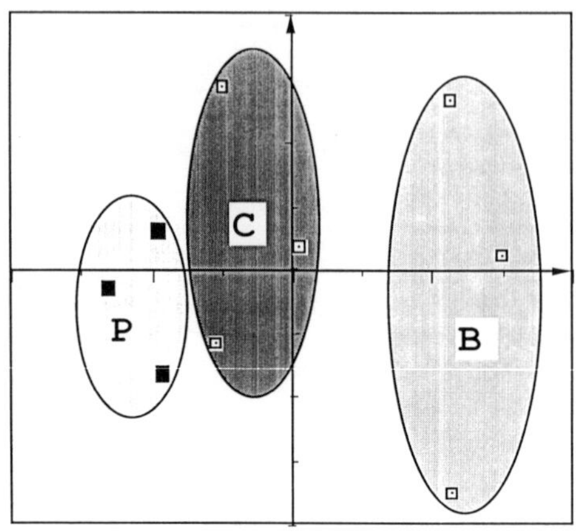

fig 12: Discriminant Factorial Analysis: Discrimination of cross-link states with unfreezable water (DSC) and R⁺ (NMR)

REFERENCES :

Bailey A.J., **Peach M.C.** and Fowler L.J. (1970), Biochem. J. 117, 819-319.
Bailey A.J. (1985), J. of Animal Sci. 60 (6), 1580-1587
Chein J.C.W. (1975), J. Macro mol. Sci. Rev. Macromol. Chem. 12, 1-80.
Goldman M. and **Shen L.** (1966), Phys. Rev. 144, 321-331.
Hazlewood C.F., **Egan T.F.** and **Rorschach H.E.** (1988), In:Water and ions in Biological Systems. Proceed. 4th Intern. Conf. (P. Läuger, L. Packer and V.Vasilescu, Eds)pp. 367-371, (Birkhauser congress reports, Life Sci vol 1)
Kopp J. (1971), Bull.Techn. CRZV Theix, INRA 5, 27, 47-55.
Kopp J. and **Bonnet M.** (1982), Bull.Techn. CRZV Theix, INRA 48, 34-37
Kopp J. and **Bonnet M.** (1986), Adv. in Meat Res. 4, 163-185
Kopp J., **Bonnet M.** and **Renou J.P.** (1989), Collagen Rel. Res. (in press).
Ledward D.A. (1984), J. Sci. Food Agric. 35, 1262
Mrevlishvili G.M. and **Privalov P.L.** (1969), In: Water in biological systems (Kayushin Ed.), Consult. Bureau., New York, pp. 63-67.
Pineri M.H., **Escoube M.** and **Roche G.** (1978), Biopolymers 17, 2799-2815.
Susi H., **Ard J.S.** and **Caroll R.J.** (1978), Biopolymers 10, 1597-1604.
Renou J.P., **Alizori J.**, **Dohri M.** and **Robert H.** (1983), J. Biochem. Biophys. Meth. 7, 91-99.

PROTON CONDUCTIVITY IN HYDRATED PROTEINS. EVIDENCE FOR PERCOLATION

G. Careri, G. Consolini, and F. Bruni

Dipartimento di Fisica, Universita' di Roma I, 00185 Roma, ITALIA

ABSTRACT

Proton conductivity in hydrated lysozyme powders displays a 2-dimensional percolation character above a critical moisture threshold, which coincides with the onset of catalytic activity. Here we shall present the behavior of this protein at low temperature, and the generalization to a complex system of proteins. Dielectric data on hydrated protein powders have been obtained in the 10 KHz - 1 MHz range from room down to liquid nitrogen temperatures.

At the lowest temperature investigated the conductivity is electronic, while with increasing temperature a protonic contribution becomes superimposed and grows to reach an Arrhenius law only at room temperature. As shown by deuterium substitution, in the intermediate temperature range the conductivity is likely due to proton tunneling. Quite similar results have been obtained in powdered samples of corn embryos, an intact system of several enzymes imbedded in membranes.

PERCOLATION MODEL

A general statistical-physical approach, called the percolation model, has been shown to be applicable to a wide range of processes where spatially random events and topological disorder are of intrinsic importance. A typical physical application of the percolation theory (Zallen, 1983), is to the electrical conductivity of a network of conducting and non-conducting elements. One of the most appealing aspects of the percolation process is the presence of a sharp transition, where <u>long-range connectivity</u> among the elements of a system <u>suddenly appear</u> at a critical concentration of the carriers. My aim here is to show that this model can successfully describe the emergence of biological function in some very simple bio-materials. This is because biological systems are often disordered at microscopic scale; then long-range connectivity between subunits must be established according to statistical laws, and the presence of a threshold should be expected according to the percolation model.

In the typical example of a network of conducting and non conducting elements, percolation theory predicts the critical concentration P_C of the conducting elements for the onset of the percolative process, and the critical exponent t for the conductivity σ dependence on P above this threshold

$$\sigma = \sigma_C + k(P-P_C)^t \tag{1}$$

In eq. (1) the kinetic coefficient k depends on the specific process in question, while P_C and t are universal quantities which are only dependent from the dimensionality D of the system . For a given system hydrated with H_2O or D_2O, if the charge carriers are protons we must expect

$$(k_{H_2O}/k_{D_2O}) = 2^{1/2} \tag{2}$$

Thus percolation theory allow specific predictions to be made on the nature of the moving changes and on the dimensionality of the conduction process.

PROTONIC PERCOLATION IN LYSOZIME POWDERS

Lysozyme is a comparatively simple enzyme, and almost everything is known about its hydrated powders, thanks to I.R. spectroscopy, E.P.R.

relaxation, heat capacity and other thermodynamic and dynamic properties, and especially the enzyme activity towards appropriate substrates(Rupley et al.,1983). We may summarize these data by saying that the hydration-stepwise process consists of three well-defined stages: 1) from 0 to about 60 H_2O molecules/macromolecule, dominated by the interaction of water with the charged groups of the protein; 2) from 60 to about 220 H_2O molecules/macromolecule, where some major changes in surface water arrangements take place; 3) from 220 to about 300 or more H_2O molecules/macromolecule, where the enzymatic activity starts and grows with increasing hydration, together with the condensation of water molecules onto weakly interacting unfilled patches of surface where the molecules are in rapid motion. It is important to note that no structural transitions in the adsorbed water or in the protein itself have been detected in the range from 60 to 220 H_2O molecules/macromolecule (or hydration h included between 0.07 and 0.25 g H_2O/g dry weight).

Recent work by our group has shown that powders of lysozyme at low hydration display protonic conductivity (Careri et al., 1985) and that the conduction process follows the percolation model (Careri et al., 1986). In this picture, the conductivity reflects motion of protons along threads of hydrogen-bonded water molecules adsorbed on the surface of the macromolecule, with long-range proton movement developing along with the extended network at the percolation threshold. In our more recent work we have been able to detect the critical exponents of this process (Careri et al., 1988).

In native Lysozyme powders the dielectric capacitance displayed a sharp increase at a water content threshold h=0.150 $\pm$ 0.016 g/g, followed by saturation at increasing hydration (Careri et al., 1986). From the capacitance data at different frequencies one can derive the d.c. conductivity σ, which displays a similar sharp increase. Since the hydration of one monolayer is h=0.38 $\pm$ 10% g/g, the experimental volume ratio for surface percolation is 0.40 $\pm$ 10%, a value very close to the 0.45 $\pm$ 0.03 predicted by theory (Zallen, 1983). Notice that for three-dimensional networks, regardless of their structure, the conduction threshold predicted by theory is 0.16 $\pm$ 0.02, and this rules out connectivity through the protein interior, where water molecules are known to be very sparse. Moreover, the threshold h_c was found to be constant from pH 3 to pH 8, indicating that the local geography of water clusters about ionizable sites of the protein surface is not of primary importance. Thus only the number of water molecules acting as interconnected conductivity sites is relevant; and as a matter of fact, the same threshold is found for both H_2O- and D_2O-hydrated samples.

From equation (1) one can easily derive that above the threshold the conductivity σ must follow the power law

$$\sigma(h) - \sigma(h_c) = k (h - h_c)^t$$

where t depends on the dimensionality of the system. Result of this analysis are in very good agreement with the theoretical prediction for a 2D conduction process (Careri et al., 1988). The previous analysis of the dielectric data reached independently the same conclusion about the dimensionality of the percolation from the close agreement between the measured value of h_c and the prediction from theory for a surface process (Careri et al., 1988). The dielectric response at hydration levels near h_c reflects protonic conduction over pathlenghts of the order of the diameter of a single macromolecule.

For lysozyme-saccharide complexes a higher value of the percolation threshold has been found, suggesting that the presence of a "foreign body", where the water bridges may not be favorable for proton transfer, must affect the long-range connectivity on the protein surface. This hydration level, $h_c=0.25$, is so close to the critical level for the onset of enzymatic activity in Lysozyme powders (Rupley et al., 1983) that it suggests protonic percolation is involved in Lysozyme catalysis. The value of t found for the saccharide complex suggests that the protonic conduction remains a surface process. This observation is in agreement with the suggestion that preferred paths of proton movement pass through the active site; substrate would be expected to block these without changing the surface character of the percolation.

QUANTUM TUNNELLING

Here let me report some preliminary results (Careri et al., to be published) on the low temperature conductivity of hydrated lysozyme powders, to investigate the possible occurence of proton quantum tunnelling in extended networks of hydrogen bonded water molecules. In this work our capacitor was cooled to liquid N_2 temperature by conventional cryogenics, and the dielectric data from 10 kHz to 1MHz have been recorded while heating at a rate of about 1 degree per minute. A typical run lasted about 5 hours and included about 300 conductivity vs temperature data. Native lysozyme was at pH 7.

We first consider the Arrhenius plot in the high temperature range. The isotopic factor in the high temperature limit is found close to the classical value $2^{1/2}$, in agreement with previous work, as it can be seen by comparing the reduced conductivities of H_2O-versus D_2O-hydrated samples at nearly the same hydration levels. We conclude that the room temperature conductivity of this system obeys the transition state theory, and displays the dependence on the hydration level typical of most biomaterials.

Next we consider the temperature region where tunnelling may prevail. A general theory of a quantum system which can tunnell out of a metastable state and which interacts with an environment at temperature T has been produced by Grabert, Weiss and Hanggi (GWH) (1984), with the finding that for damping of arbitrary strenght the tunnelling decay rate always matches smoothly with the Arrhenius factor, and that heat enhances the tunnelling probability at T=0 K by a factor exp [A(T)]. For undamped system A(T) is exponentially small, whereas for a dissipative system A(T) grows algebrically with temperature. Of particular interest here is the case of tunnelling centers in solids, where A(T) increases proportional to T^4 at low temperature. We have plotted the conductivity data versus T^4, and we have found that a remarkably simple description can be offered as follows. In this T^4 plot, the conductivity data ln $\sigma(h,T)$ can be fitted by straight lines originated near T = 0 K, and after a break adjust themself on parallel lines. This break is slightly hydration dependent, and very close to the glass transition temperatures reported by other authors in proteins (Goldanskii et al., 1986). By comparing data of H_2O- and D_2O samples at nearly the same hydration level in a T^4 plot, the slope of H_2O- is found to be higher than that of D_2O-hydrated ones, as one would expect for a tunnelling phenomenon.

These low temperature studies have been extended to powdered fragments of Corn embryos, a viable system already studied at room temperature (Bruni et al., 1989). Here the protonic conductivity follows the percolation model as well, with a critical exponent t=1.23 typical for a two dimensional process, and with an hydration threshold close to the onset of the hydration-induced respiration in intact seeds. At low hydration, in the temperature range between 180 K and 280 K, the conductivity displays a behaviour similar to the low temperature region, a protein matrix phase transition in the intermediate temperature range, and a limiting Arrhenius low at high temperature (Careri et al., to be published).

<u>REFERENCES</u>

Bruni, F., Careri, G., and Leopold, A.C. (1989) Phys. Rev. A. 40 (in press).

Careri, G., Geraci, M., Giansanti, A., and Rupley, J.A. (1985) Proc. Natl. Acad. Sci. U.S.A. <u>82</u>, 5342.

Careri, G., Giansanti, A., and Rupley, J.A. (1986) Proc. Natl. Acad. Sci. U.S.A. <u>83</u>, 6810.

Careri, G., Giansanti, A., and Rupley, J.A. (1988) Phys. Rev. A <u>37</u>, 2703.

Goldanskii, V.I., Krupyanskii, Y.F., and Flenrov, V.N. (1986) Physica Scripta <u>33</u>, 527.

Grabert, H.V., Weiss, U., and Hanggi, P. (1984) Phys. Rev. Lett. <u>52</u>, 2193.

Rupley, J.A., Gratton, E., and Careri, G. (1983) Trends in Biochem. Sciences <u>8</u>, 18.

Zallen, R. (1983) The Physics of Amorphous Solids. John Wiley and Sons, New York.

A NON-TRADITIONAL ROLE FOR WATER DURING ELECTRON TRANSFER
BY CYTOCHROME C OXIDASE

JACK A. KORNBLATT AND GASTON HUI BON HOA
DEPT. OF BIOLOGY, CONCORDIA UNIV.,
1455 de Maisonneuve
MONTREAL , Qc. H3G 1M8
AND
U 310, INSERM
INST. DE BIOL. PHYSICO-CHIMIQUE,
13, rue Pierre et Marie Curie
75005 PARIS

Summary
   Cytochrome c oxidase has been perturbed using a combiation of high hydrostatic pressure and high osmotic pressure. An ititial step that includes the entry of water has been identified in the mechanism. Approximately 10 water molecules may be involved. Subsequent steps include conformational changes and water exit.

THE CYTOCHROME C OXIDASE REACTION IS SHOWN BELOW; THE REACTION HAS NOT BEEN BALANCED

$$e^- + a^3Cu_A(II)Cu_B(II)a_3{}^3 \text{--------> ------>}\ a^2Cu_A(I)Cu_B(II)a_3{}^3$$

"CONFORMATIONAL CHANGES"
$$\text{-------------------> ----------------------->}\ a^2Cu_A(I)Cu_B(II)a_3{}^3 \text{------>}$$

$$a^3 Cu_A(II)Cu_B(I)a_3{}^2 \xrightarrow[H^+]{O_2} a^3Cu_A(II)Cu_B(II)a_3{}^3 + H_2O$$

THE STRUCTURE OF THE OXIDASE AND THE REACTION HAVE BEEN PERTURBED USING HYDROSTATIC AND OSMOTIC PRESSURE. THE PERTURBATIONS HAVE BEEN EVALUATED ON THE BASIS OF SPECTRAL CHANGES EXHIBITED BY THE OXIDASE.THE RATIONALE FOR THE EXPERIMENTS IS BASED ON THE IDEA THAT HYDROSTATIC PRESSURE (Ph) AND OSMOTIC PRESSURE (Po)
WILL INFLUENCE DIFFERENT ASPECTS OF THE REACTION SHOWN
ABOVE.
   Ph WILL DRIVE THE REACTION TOWARDS THOSE FORMS IN WHICH THE SYSTEM OCCUPIES THE SMALLEST VOLUME.Po WILL DRIVE THE REACTION TOWARDS THOSE FORMS IN WHICH OSMOTICALLY SENSITIVE COMPARTMENTS ARE IN THE DEHYDRATED STATE.

   THE BASIC EQUATIONS USED IN THE EVALUATION OF THE EFFECTS OF THE TWO PRESSURES ARE:

$$\Delta V\ =\ RTd(lnK)/dP$$

AND

$$\Delta V_c + \Delta V_o = RTd(\ln K)/dP_h$$

$$\Delta V_o = RTd(\ln K)/dP_o$$

WHERE THE SUBSCRIPTS c, o, and h STAND FOR CONFORMATION, OSMOTIC AND HYDROSTATIC.

<u>EXPERIMENTAL:</u>

THE CONTROLS:

GLYCEROL HAS LITTLE EFFECT ON THE OXIDIZED AND REDUCED CYTOCHROME OXIDASE AT CONCENTRATIONS UP TO 60% (v/v).AT 75% AND 90% GLYCEROL, THERE IS A NOTICABLE SHIFT OFTHE SPECTRA TOWARDS LOW-SPIN FORMS FOR BOTH THE TOTALLY OXIDIZED AND TOTALLY REDUCED SPECIES.NOTWITHSTANDING THE LARGE SPECTRAL SHIFTS, THERE DOES NOT APPEAR TO BE A MAJOR PERTURBATION OF THE PROTEIN AS JUDGED BY THE RATIO OF THE ALPHA PEAK (600 nm) TO THE SORET PEAK (420 nm AND 445 nm). SIMILARLY, THE COMBINED EFFECTS OF GLYCEROL AND HYDROSTATIC PRESSURE ON THE SPECTRA OF THE TOTALLY OXIDIZED OR REDUCED OXIDASE ARE LIKE THOSE OF PRESSURE IN THE ABSENCE OF GLYCEROL. THE EFFECTS OF HYDROSTATIC PRESSURE ON THE OXIDASE IN AQUEOUS SOLUTION HAVE BEEN PUBLISHED. BELOW WE SHOW THE EFFECTS OF HYDROSTATIC PRESSURE ON THE OXIDASE IN 50% GLYCEROL.(see figure 1)

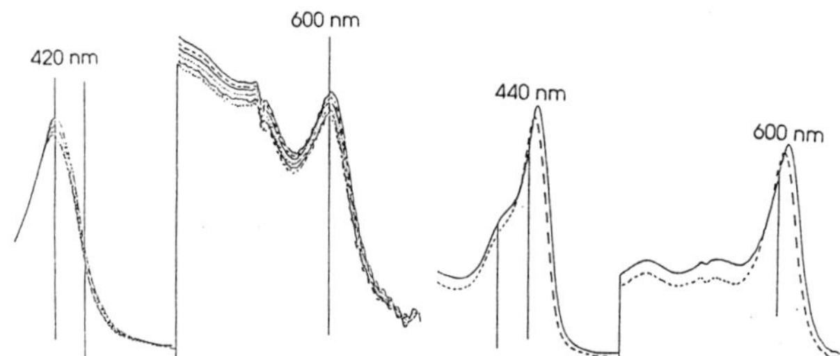

HYDROSTATIC PRESSURE INDUCES A 3 nm RED SHIFT IN THE SORET PEAK OF THE OXIDIZED PROTIEN AND A 3 nm SHIFT IN THE ALPHA AND SORET PEAKS OF THE REDUCED PROTEIN.
RED SHIFTS RESULTING FROM HIGH HYDROSTATIC PRESSURE ARE FREQUENTLY OBSERVED. WE HAVE NOT BEEN ABLE TO CORRELATE THE EXTENT OF THE SHIFT IN THE REDUCED PROTEIN WITH ANY KNOWN PROPERTY OF THE OXIDASE.

<u>THE MAJOR FINDINGS:</u>

THE COMBINED EFFECTS OF PRESSURE ON THE CYTOCHROME OXIDASE DURING TURNOVER. AT HIGH HYDROSTATIC PRESSURE, IN THE ABSENCE OF OSMOTIC PERTURBANTS, HYDROSTATIC PRESSURE INTRODUCES A BLOCK IN ELECTRON TRANSPORT SUCH THAT THE MAJOR SPECIES PRESENT IS

$$a^2Cu_A(I)Cu_B(II)a_3^3$$

THIS RESULT HAS ALREADY BEEN PUBLISHED. WHEN THE OXIDASE IN GLYCEROL CONTAINING SOLUTIONS IS PERTURBED WITH HYDROSTATIC PRESSURE A <u>SIMILAR</u> BLOCK IS OBSERVED. THE SPECTRA OF THE OXIDASE, DURING TURNOVER, IN GLYCEROL CONTAINING SOLUTION AND UNDER VARYING HYDROSTATIC PRESSURE, IS SHOWN IN THE FIGURE BELOW . (figure 2)

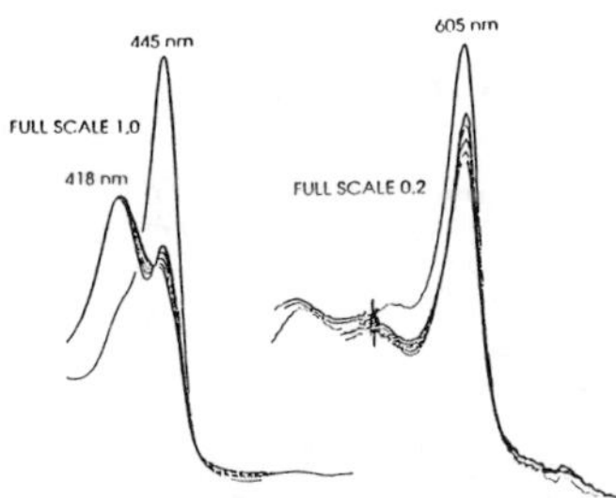

FROM THE SPECTRA, SPECIFICALLY THE ALPHA PEAK, WE WERE ABLE TO DETERMINE THE EXTENT OF REDUCTION OF CYTOCHROME a   AT EACH PRESSURE AND THIS WAS RELATED TO A QUANTITY

"Kapp"

WHICH ALLOWED US TO CALCULATE THE VOLUME CHANGE ASSOCIATED WITH THE BLOCK AT EACH GLYCEROL CONCENTRATION. A REPRESENTATIVE PLOT OF ln(Kapp) vs PRESSURE IS SHOWN IN FIGURE 3.

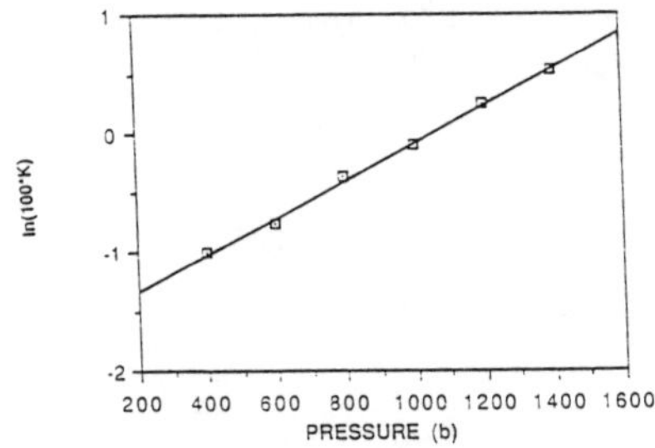

THE VOLUME CHANGES VARIED AS A FUNCTION OF GLYCEROL CONCENTRATION AS SHOWN BELOW. (figure 4)

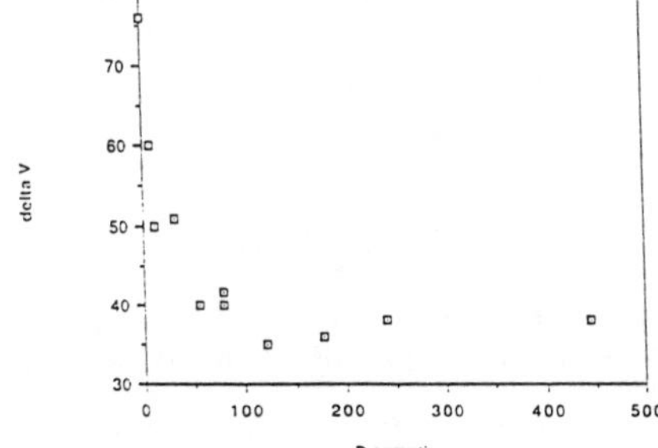

MORE IMPORTANTLY, THE VALUE OF "Kapp" VARIED SYSTEMATICALLY AS A FUNCTION OF GLYCEROL CONCENTRATION AND THEREFORE AS A FUNCTION OF Po.
THIS VARIATION IS SHOWN IN THE FIGURE BELOW.(figure5)

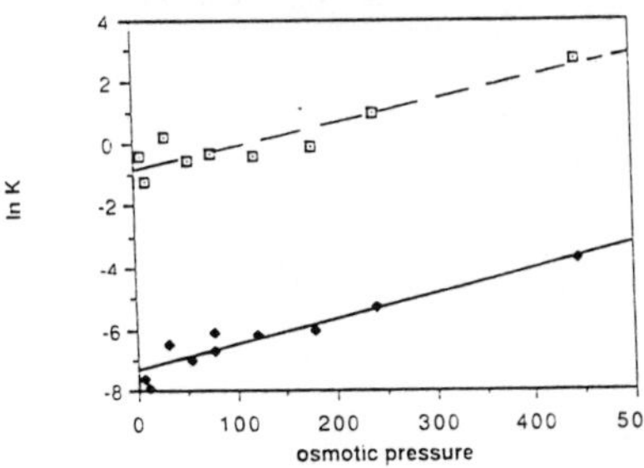

THE TWO FIGURES SHOWN ABOVE CORRESPOND TO THE EXPERIMENTAL FACTS.
OUR FIRST ASSUMPTION: THE DATA OF THE ABOVE FIGURE CAN BE EVALUATED TO YIELD A QUANTITY

$$\Delta V_O$$

WHICH IS RELATED TO THE AMOUNT OF SOLVENT WATER WHICH MUST ENTER THE OXIDASE FOR ELECTRON TRANSFER FROM

$$a^2Cu_A(I) \qquad to \qquad Cu_B(II)a_3^3$$

TO OCCUR. THE ENTRY OF THIS WATER IS DISFAVOURED BY HIGH OSMOTIC PRESSURE AND FAVOURED BY HIGH HYDROSTATIC PRESSURE. THE VALUE OF ΔV_O IS 160 mL mol^{-1} AT 1b.
OUR SECOND ASSUMPTION IS THAT THE QUANTITIES P_O AND P_h ARE RELATED AND THAT Ph CAN PROBE ALL PROCESSES FOR WHICH ΔV IS FINITE.

FOR THE OXIDASE THIS CORRESPONDS TO ALL VOLUME CHANGES ASSOCIATED WITH CONFORMATIONAL CHANGES AS WELL AS THOSE ASSOCIATED WITH SOLVENT MOVEMENT.

$$\Delta V_h = \Delta V_C \;+\; \Delta V_O$$

P_O, ON THE OTHER HAND, CAN PROBE PROCESSES IN WHICH OSMOTICALLY SENSITIVE COMPARTMENTS ARE EXPOSED DURING TURNOVER.

$$\Delta V_O = \Delta V_O$$

IF OUR FIRST ASSUMPTION IS CORRECT, ie

$$\Delta V_O \quad = \quad RTd(lnK)/dP_O$$

AND IF OUR SECOND ASSUMPTION IS ALSO CORRECT, ie

$$\Delta V_h \quad = \quad \Delta V_C \quad + \quad \Delta V_O \qquad THEN$$

$$-80 \; mLmol^{-1} = \Delta V \; + \; -160 \; mLmol^{-1} \qquad AND$$

$$\Delta V_{con} = \; +80 \; mLmol^{-1}$$

THE VOLUME CHANGE ASSOCIATED WITH THE CONFORMATIONAL CHANGE HAS A POSITIVE SIGN.
IN CONCLUSION, WE FEEL THAT WE HAVE PROBABLY IDENTIFIED A STEP IN THE CYTOCHROME OXIDASE REACTION THAT INVOLVES A SUBSTANTIAL AMOUNT OF WATER
FLOW. WHEN WE INTEGRATE THIS INFORMATION WITH OUR ORIGINAL CATALYTIC SCHEME WE ARRIVE AT THE FOLLOWING:

$$e^- + a^3Cu_A(II)Cu_B(II)a_3^3 \; \text{--------------------->} \; a^2Cu_A(I)Cu_B(II)a_3^3$$

WATER ENTERS, ΔV, = 160 mLmol^{-1}
$\text{---} \rightarrow \; a^2Cu_A(I)Cu_B(II)a_3^3$
  INHIBITED BY HIGH P_O

  CONFORMATIONAL CHANGE
    WATER EXITS, ΔV_{net} = 80 mLmol^{-1}
$\text{---} \rightarrow a^3 Cu_A(II)Cu_B(I)a_3^2$
  INHIBITED BY HIGH P_h

$$O_2$$

$$a^3 Cu_A(II)Cu_B(I)a_3^2 \; \text{--------} > \; a^3Cu_A(II)Cu_B(II)a_3^3 \; + \; H_2O$$
$$H^+$$

THE REMAINING CONFORMATIONAL CHANGE OR CHANGES MAY LOCALIZED IN THE FIRST STEP OR MAY BE DISTRIBUTED AMONGST OTHER STEPS. WORK IS IN PROGRESS TO IDENTIFY WHICH STEP OR STEPS.

WATER, IONS AND MEMBRANES

ION TRANSLOCATION BY THE NA,K-PUMP: CORRELATION BETWEEN ELECTROGENIC EVENTS AND CONFORMATIONAL TRANSITIONS

H.-J. Apell, R. Borlinghaus, P. Läuger, W. Stürmer and I. Wuddel

Department of Biology, University of Konstanz, D-7750 Konstanz, F.R.G.

SUMMARY

The electrogenic properties of the Na,K-ATPase can be studied by correlating transient electrical events in the pump molecule with optically-detected conformational changes. Membrane fragments containing a high density of oriented ATPase molecules can be bound to a planar bilayer acting as a capacitive electrode. Flash-induced ATP-release in the solution from "caged" ATP elicits transient electrical signals in the external measuring circuit, reflecting charge movements in the pump molecule. In parallel experiments, optical signals resulting from conformational transitions were recorded from fluorescence-labeled Na,K-ATPase. From the comparison of electrical and optical transients recorded under a variety of conditions, information on the nature of the charge-translocating steps in the pumping cycle can be obtained.

INTRODUCTION

The sodium-potassium pump in the plasma membrane of mammalian cells carries out uphill transport of sodium and potassium ions at the expense of free energy of ATP hydrolysis (Skou, 1975; Glynn, 1985; Andersen and Jørgensen, 1988; Apell, 1989). Spectroscopic and other studies indicate that the enzyme can assume two principal conformations designated E_1 and E_2. Form E_1 has the ion-binding sites facing the cytoplasm and is stabilized by Na^+, form

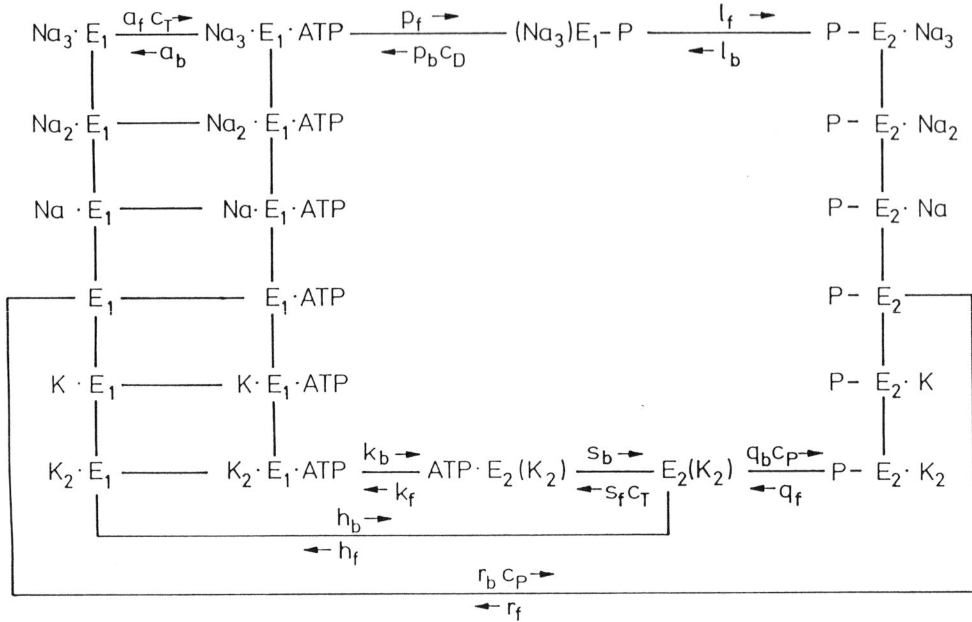

<u>Fig. 1</u> Post-Albers scheme for the pumping cycle of Na,K-ATPase (Cantley et al. 1984). E_1 and E_2 are conformations of the enzyme with ion binding sites exposed to the cytoplasm and the extracellular medium, respectively. In the "occluded" states $(Na_3)E_1$-P and $E_2(K_2)$ the bound ions are unable to exchange with the aqueous phase. Dashes indicate covalent bonds and dots indicate noncovalent bonds. a_f, h_f, k_f, ... and a_b, h_b, k_b, ... are rate constants for transitions in forward and backward direction, respectively. c_T, c_D and c_P are cytoplasmic concentrations of ATP, ADP and P_i (inorganic phosphate). The rate constants a_f and a_b are assumed to be the same for all transitions $Na_i \cdot E_1 \leftrightarrow Na_i \cdot E_1 \cdot ATP$ and $K_j \cdot E_1 \leftrightarrow K_j \cdot E_1 \cdot ATP$ (i=0,1,2,3; j=1,2).

E_2 has the ion-binding sites facing the extracellular medium and is stabilized by K^+. From enzymatic and transport studies the reaction cycle represented in Fig. 1 has been proposed (Cantley et al., 1984). When the protein is phosphorylated in state E_1 by ATP, Na^+ becomes "occluded", i.e., trapped inside the protein $(Na_3 \cdot E_1 \cdot ATP \rightarrow (Na_3)E_1$-P). After transition to conformation E_2, Na^+ is released and K^+ is bound. This leads to dephosphorylation of the protein and occlusion of K^+. The original state is restored by transition to conformation E_1 and release of K^+ to the cytoplasmic side.

Since (under normal conditions) in a single turnover three Na^+ ions are moved outward and two K^+ inward, the transport process is associated with the translocation of net charge. The electrogenic nature of the Na,K-pump has interesting consequences. The pump acts as a current generator and contributes to the membrane potential of cells. Furthermore, the transport rate becomes a function of transmembrane voltage. Of particular interest is the question in which step (or steps) of the transport cycle charge is translocated. This problem may be studied by recording transient, pump-generated currents after an ATP-concentration jump and by correlating the electrical signals with optical signals observed with fluorescence-labeled Na,K-ATPase.

Fast charge translocations elicited by ATP-concentration jumps

Information on the nature of charge-carrying steps and on the kinetic parameters of the pumping cycle may be obtained from experiments in which nonstationary pump-currents are induced by a sudden change of ATP concentration (Fendler et al., 1985; Borlinghaus et al., 1987; Apell et al., 1987). For the measurement of the transient current signals, flat membrane-fragments rich in Na,K-ATPase are bound to a planar lipid bilayer acting as a capacitive electrode (Fig. 2).

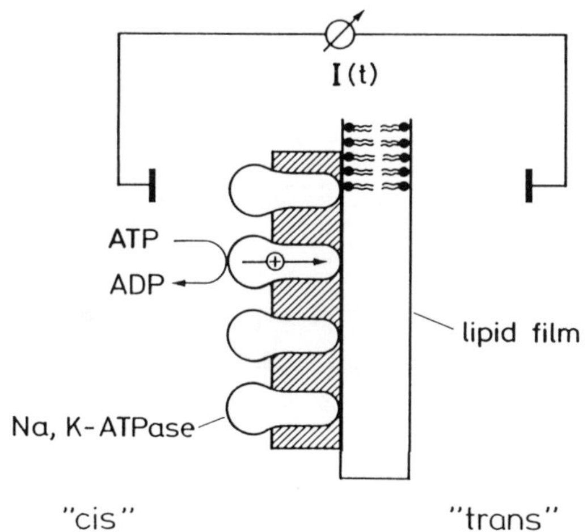

Fig. 2 Planar lipid bilayer with bound Na,K-ATPase membrane fragments. When a suspension of flat membrane fragments is added to the aqueous medium, fragments become bound to the lipid bilayer, some with the cytoplasmic side facing the solution. The diameter of the fragments is in the range of $0.1-1 \, \mu m$; the density of oriented Na,K-ATPase molecules in the fragment is about $10^3-10^4 \, \mu m^{-2}$. Flash-induced release of ATP from a photolabile ATP-derivative ("caged" ATP) elicits a transient current I(t) in the external measuring circuit.

The membrane preparation which is obtained by dodecylsulfate extraction of kidney microsomes (Jørgensen, 1974) consists of flat membrane sheets 0.1-1 μm in diameter containing oriented Na,K-ATPase molecules with a density of several thousand per μm^2 (Deguchi et al., 1977). In the aqueous phase which is in contact with the bound membrane sheets, ATP is released within milliseconds from an inactive, photolabile derivative ("caged" ATP) by an intense flash of light (McGray et al., 1980). After the ATP-concentration jump which leads to a (nearly) simultaneous activation of many pump molecules, transient current and voltage signals can be recorded in the electrical circuit connecting the aqueous phases adjacent to the lipid bilayer (Fig. 2). At pH 7.0 ATP is liberated from "caged" ATP with a time constant of 4.6 msec (McCray et al., 1980). With a starting concentration of caged ATP of 0.5 mM, the concentration of released ATP after a single flash is typically ~50 μM.

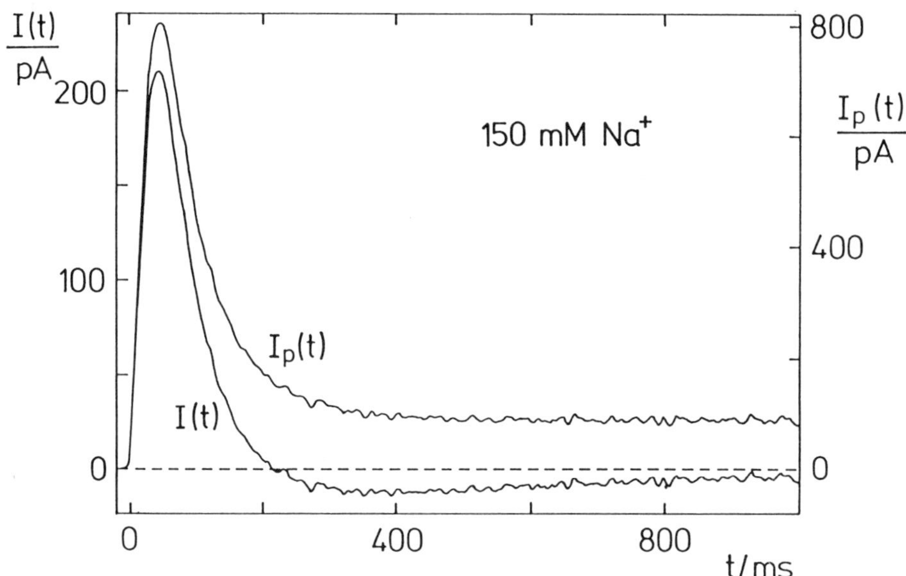

Fig. 3 Current signal I(t) in the absence of K$^+$ from a lipid bilayer with bound Na,K-ATPase membrane fragments (Fig. 2). At time t=0, about 20 μM ATP was released from caged ATP by a 40 μs light flash. I(t) was computed from the recorded voltage signal V(t) according to $I(t)=-AC_f dV/dt$. The area A of the black film (specific capacitance $C_f \approx 0.37$ μF/cm^2) was determined by an eye-piece micrometer (A ≈ 0.60 mm^2). The aqueous solutions contained 150 mM NaCl, 30 mM imidazole, pH 7.2 and 10 mM MgCl$_2$. The temperature was 20 ^{O}C. Membrane fragments (40 μg/ml) and 240 μM "caged" ATP were added to the "cis" side (Fig. 2) 20 min prior to the experiment. The voltage signal was recorded at a bandwidth of 1 kHz. The intrinsic pump current $I_p(t)$ was evaluated from I(t) according to Eq. (2) of Borlinghaus et al. (1987).

<u>Electrical signals in the presence of Na$^+$</u>

Membrane fragments were added together with 240 μM caged ATP to one side of a planar lipid bilayer. After a waiting time of 20-30 min, which is probably required for adsorption of membrane fragments, the photoresponse of the system was fully developed. A light pulse of 40 μs duration which liberated about 20 μM ATP on the "cis" side of the bilayer (Fig. 2) elicited a transient electric current I (Fig. 3). In these experiments the planar lipid bilayer acts as a capacitive element which couples electrical events in the protein layer to the external measuring circuit (Borlinghaus et al., 1987). In the experiment represented in Fig. 3, the aqueous solutions contained Na$^+$, but no K$^+$. The sign of the early phase of the current corresponds to a movement of positive charge from the solution toward the lipid bilayer. The transient current is generated by those membrane fragments which are bound to the bilayer with the cytoplasmic side facing the aqueous medium (Borlinghaus et al., 1987).

Charge translocation after activation of the pump can be detected either as a current I(t) under short-circuit conditions (Fig. 2), or as a voltage V(t). It has been demonstrated previously (Borlinghaus et al., 1987) that voltage and current signals are strictly correlated according to the relation I(t)=-AC$_f$dV/dt, where A and C$_f$ are the area and the specific capacitance of the lipid film, respectively. Since a measurement of V(t) yields a better signal-to-noise ratio, in most experiments voltage signals were recorded. V(t) was subsequently converted into a current I(t) by numerical differentiation.
From the current signal in the external measuring circuit, I(t), the intrinsic pump current $I_p(t)$ can be evaluated. $I_p(t)$ is the current which would be observed in a fictitious experiment in which a continuous layer of membrane fragments is directly interposed (without supporting bilayer) between the aqueous solutions. $I_p(t)$ and I(t) are connected by the circuit parameters of the compound membrane system which may be estimated from the shape of V(t) at long times (Borlinghaus et al., 1987). From Fig. 3 it is seen that $I_p(t)$ and I(t) differ only at large values of t. In particular, the negative phase of I(t) which results from backflow of charge across the shunt conductance of the membrane fragments is not present in $I_p(t)$. Instead, $I_p(t)$ approaches a small quasistationary current I_p^∞ at long times which results

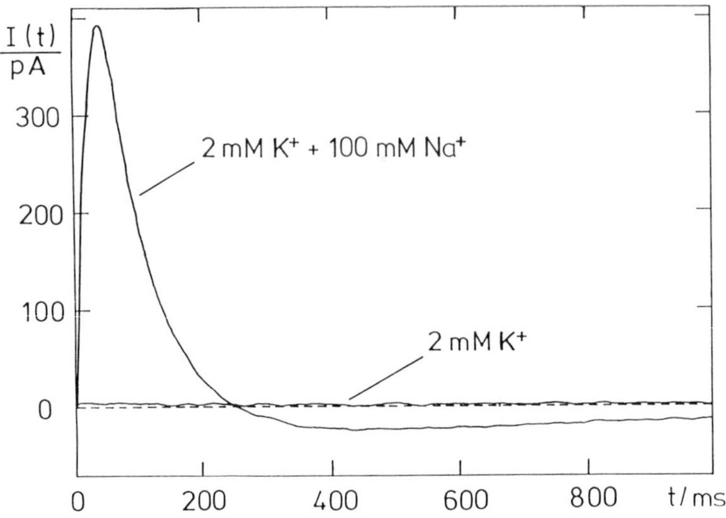

__Fig. 4__ Electrical transients in the presence of 2mM K$^+$ alone and after addition of 100 mM Na$^+$. The transient was measured as a voltage signal and was subsequently transformed into a current I(t) by digital differentiation. T=25.5 $^{\circ}$C. The other experimental conditions were the same as in Fig 3.

from pump molecules undergoing the transition P-E$_2$ → E$_1$ and reentering the cycle again (Apell et al., 1987). The small amplitude of I$_p^\infty$ is consistent with the observation that the rate of the transition P-E$_2$ → E$_1$ is extremely low (1-5 s^{-1} at 20 $^{\circ}$C) in the absence of K$^+$ (Glynn & Karlish, 1976).

Electrical signals in the presence of K$^+$

When the medium contained 2 mM K$^+$, but no Na$^+$, the transient electric current following an ATP-concentration jump was virtually zero (Fig. 4). As seen from Fig. 4, subsequent addition of 100 mM Na$^+$ and photochemical ATP-release gave rise to a large current signal. Similar results were obtained when the K$^+$ concentration was varied in the range between 1 and 10 mM.

CONFORMATIONAL TRANSITIONS STUDIED BY FLUORESCENCE MEASUREMENTS

The kinetics of conformational transitions of the pump may be studied by time-resolved fluorescence measurements, using fluorescent probes bound to the protein (Karlish et al., 1978). An important problem in the understanding of the kinetic behaviour of the Na,K-pump is the question, how the charge movements which give rise to the transient currents are correlated with the spectroscopically-detected conformational changes. In the following, we describe experiments in which transient fluorescence changes of 5-iodoacetamidofluorescein-labeled Na,K-ATPase (Kapakos & Steinberg, 1986; Steinberg & Karlish, 1989; Stürmer et al., 1989) have been recorded after an ATP-concentration jump. By comparison of the optical and the electrical signals valuable kinetic information can be obtained.

For high-resolution recording of time-dependent fluorescence signals the set-up represented in Fig. 5 was used (Stürmer et al., 1989). The optical cell contained Na,K-ATPase in the form of membrane fragments, labeled with 5-iodoacetamidofluorescein (5-IAF). ATP was liberated from "caged" ATP by 10 ns light-flashes generated by an excimer laser.

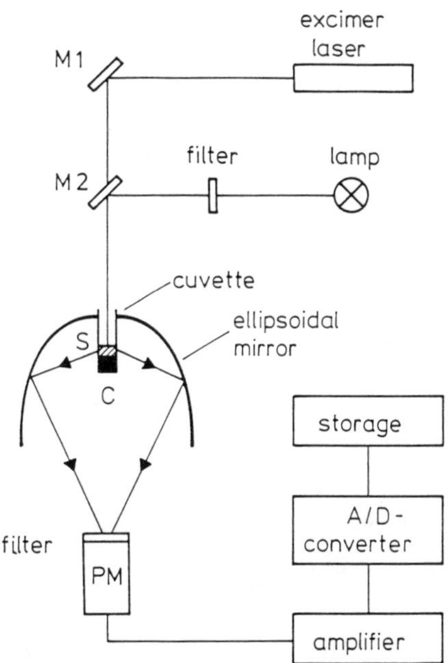

<u>Fig. 5</u> Set-up for time-resolved fluorescence measurements. The cylindrical glass cuvette containing the sample (S) is mounted on a thermostated metal console (C). Fluorescence of the sample is excited by light from a tungsten-halogen lamp. The light emitted by the sample is collected by an ellipsoidal mirror and focussed onto the cathode of the photomultiplier (PM). The light-emitting part of the cuvette is located in one focal point of the mirror and the photocathode in the other. The photomultiplier signal is amplified, digitized in an analog-to-digital converter and stored in the memory of the Compaq-386 computer. ATP is released from "caged" ATP by a 10 ns pulse from an excimer laser. The laser beam enters the cuvette through a hole in mirror M2.

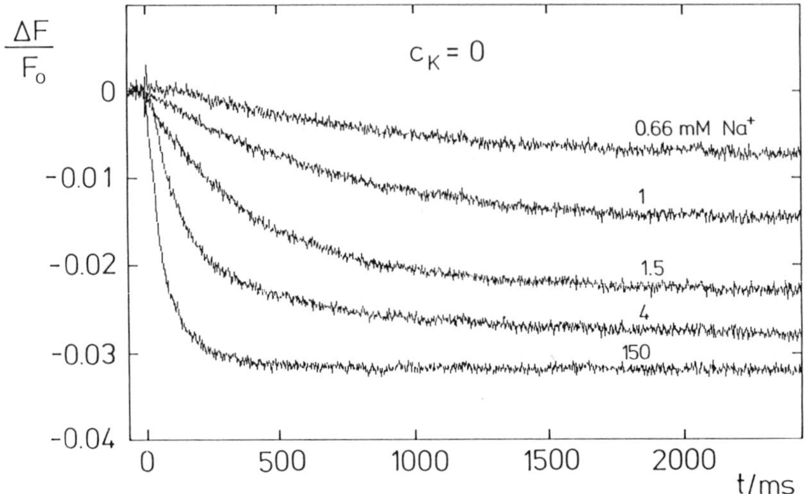

<u>Fig. 6</u> Relative fluorescence change $\Delta F/F_0$ as a function of time t for different Na^+ concentrations (0.66, 1, 1.5, 4 and 150 mM). At time t=0, about 20 μM ATP were released in the solution. The temperature was 20° C. From Stürmer et al., 1989.

<u>Fluorescence signals in the presence of Na^+</u>

The time course of the ATP-induced fluorescence change $\Delta F/F_0$ in K^+-free Na^+-solutions is shown in Fig. 6 for different Na^+ concentrations (0.66, 1, 1.5, 4 and 150 mM). $\Delta F(t)/F_0$ can be approximately fitted by a single exponential function with a time constant depending on sodium concentration. Previous studies indicate that phosphorylation by ATP in the absence of K^+ leads to the reaction sequence $Na_3 \cdot E_1 \cdot ATP \rightarrow (Na_3)E_1-P \rightarrow P-E_2 \cdot Na_3 \rightarrow P-E_2$ (Glynn, 1985). Under K^+-free conditions, $P-E_2$ is a long-lived state which only slowly (with a rate constant $r_f \approx 1-5 \ s^{-1}$) returns to state E_1. It is therefore likely to assume that the fluorescence decay in Fig. 6 reflects the $E_1 \rightarrow E_2$ conformational transition of the protein (Kapakos & Steinberg, 1986).

<u>Fluorescence signals in the presence of K^+</u>

When the medium contains K^+, but no Na^+, an ATP-concentration jump leads to a fluorescence increase, i.e., to a signal of opposite polarity compared to the signals observed with Na^+ media (Fig. 7). For t>20 ms, the time course of the fluorescence change can be approximately described by a simple

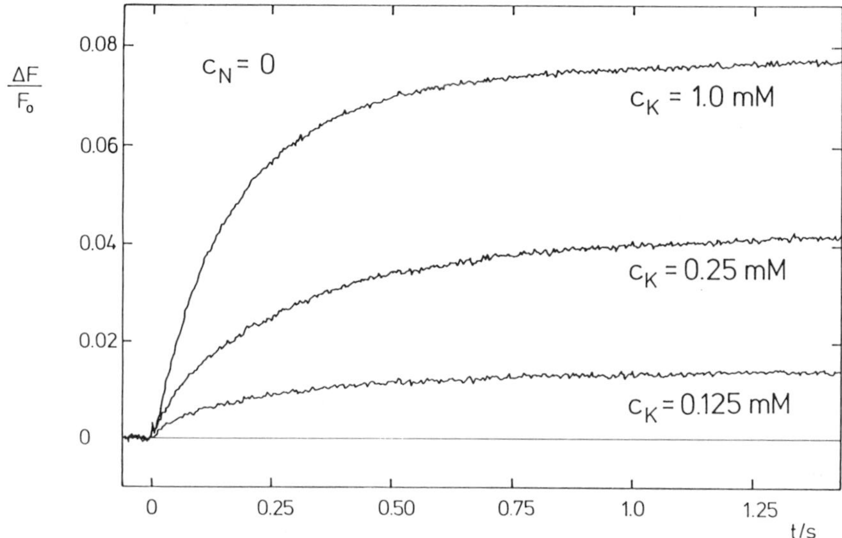

Fig. 7 Fluorescence signal $\Delta F/F_0$ in a sodium-free potassium medium. At time t=0 about 20 μM ATP were released in a solution containing 0.125 mM, 0.25 mM or 1 mM K^+; T=20 $^{\circ}$C. From Stürmer et al., 1989.

exponential relation of the form $\Delta F/F_0 = a[1-\exp(-t/\tau)]$. From other studies it is known that in the presence of millimolar concentrations of K^+ and in the absence of Na^+ and ATP, a substantial fraction of the enzyme is initially present in state $E_2(K_2)$, since the equilibrium between $K_2 \cdot E_1$ and $E_2(K_2)$ is strongly poised (about 1000-fold) towards $E_2(K_2)$ (Karlish, 1980). Binding of ATP to $E_2(K_2)$ shifts the equilibrium back to state $E_1 \cdot$ATP which gives rise to an increase of fluorescence.

Comparison of electrical and optical signals (Na^+ media)

Previous studies with chymotrypsin-modified enzyme (Borlinghaus et al., 1987) led to the conclusion that the major charge-carrying step in the sodium limb of the transport cycle is the deocclusion reaction $(Na_3)E_1-P \rightarrow P-E_2 \cdot Na_3$, followed by release of Na^+ to the extracellular medium.

The simplest situation for a comparison of optical and electrical signals is given when both the fluorescence change and the transient current result from a single reaction step $A \rightarrow B$. The fluorescence intensity F may then be represented by $F = f_A c_A + f_B c_B$, where f_A and f_B are the contributions of

A and B to F, and c_A and c_B are the concentrations. Introducing the initial fluorescence $F_0=f_Ac$, where $c=c_A+c_B$ is the total concentration, the time course of the relative fluorescence change $\Delta F/F_0\equiv(F-F_0)/F_0$ is obtained as

$$\frac{\Delta F(t)}{F_0} = \frac{1}{c}\left[\frac{f_B}{f_A} - 1\right] c_B(t) \tag{1}$$

On the other hand, the transient pump current $I_p(t)-I_p^\infty$ associated with the reaction $A \rightarrow B$ may be written as

$$I_p(t) - I_p^\infty = \alpha e_0 \frac{dc_B}{dt} \tag{2}$$

where e_0 is the elementary charge and a is a constant; I_p^∞ accounts for any quasistationary component of $I_p(t)$. According to Eq. (2), the total translocated charge $Q(t)$ is given by

$$Q(t) \equiv \int_0^t (I_p - I_p^\infty)\,dt = \alpha e_0 c_B(t) \tag{3}$$

Comparison with Eq. (1) shows that under the assumptions introduced above, $\Delta F(t)/F_0$ and $Q(t)$ should exhibit the same time course, apart from a scaling factor.

A behaviour predicted by Eqs. (1) and (3) is observed in experiments in K^+-free Na^+-media. This is shown in Fig. 8 in which $\Delta F(t)/F_0$ is plotted together with the translocated charge $Q(t)$ for $c_N=150$ mM. $Q(t)$ was obtained according to Eq. (3) as the time integral of $I_p-I_p^\infty$. It is seen that $\Delta F(t)/F_0$ and $Q(t)$ nearly coincide. This indicates that both the fluorescence change as well as the charge translocation take place in the same reaction step. As discussed above, a likely candidate for this common reaction step is the de-occlusion of sodium.

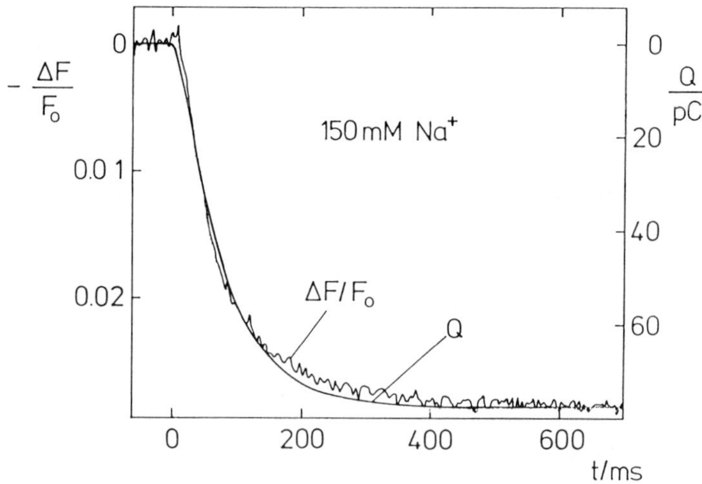

__Fig. 8__ Comparison of ATP-induced electrical and optical signals measured in parallel experiments under nearly identical conditions (150 mM Na$^+$, 30 mM imidazole, pH=7.2, 1 mM EDTA, 10 mM Mg^{++}, T=21-22 ^{o}C). $\Delta F/F_o$ is the relative fluorescence change and Q the translocated charge (Eq. 3); 1 pC= 10^{-12} coulomb.

The transient currents represented in Figs. 3 and 4 are observed under single-turnover conditions in which the pump moves through part of the reaction cycle. From the Na$^+$ experiment (Fig. 3), it is clear that the sensitivity of the current measurement is sufficient to detect charge translocation under single-turnover conditions, and therefore a current signal should have been observed in the K$^+$ experiment (Fig. 4), if the transitions $E_2(K_2)$ → $K_2 \cdot E_1$ → E_1 are associated with charge movement. In the optical experiment which was carried out under virtually identical conditions, a large fluorescence change was observed after the light flash (Fig. 7), indicating that an appreciable shift of the conformational equilibrium between $E_2(K_2)$ and $K_2 \cdot E_1$ occurs after ATP release. Comparison of the optical and electrical signals thus leads to the conclusion that the transition ATP$\cdot E_2(K_2)$ → $K_2 \cdot E_1 \cdot$ATP is electrically silent.

CONCLUSIONS

In this study we have attempted to correlate transient electrical events in the Na,K-pump with conformational transitions elicited by an ATP-concentration jump. When the medium contains Na$^+$, but no K$^+$, the fluorescence of the 5-IAF-labeled protein decreases monotonously after activation by ATP. In parallel experiments carried out under otherwise identical conditions with membrane fragments bound to a planar bilayer, a transient pump current $I_p(t)$ was observed which decayed with nearly the same time behaviour. Apart from a scaling factor, the fluorescence signal $\Delta F(t)/F_0$ was found to nearly superimpose with the time integral $Q(t)$ of I_p (Fig.8). The close correlation between $\Delta F/F_0$ and $Q(t)$ indicates that the optical and the electrical transient are governed by the same rate-limiting step. Chymotrypsin modification of the protein, which is known to block the transition from $(Na_3)E_1-P$ to $P-E_2 \cdot Na_3$, was found to eliminate both the fluorescence signal and the current transient (Borlinghaus et al., 1987; Stürmer et al., 1989). This strongly suggests that the fluorescence change as well as the charge movement take place after formation of the occluded state, i.e., in the transitions $(Na_3)E_1-P \rightarrow P-E_2 \cdot Na_3 \rightarrow P-E_2$. Assuming that release of sodium ($P-E_2 \cdot Na_3 \rightarrow P-E_2$) is fast, the experimentally observed time-constant of the fluorescence decay may be identified with the chemical relaxation time of the transition $(Na_3)E_1-P \rightleftharpoons P-E_2 \cdot Na_3$.

In experiments with Na^+-free K^+-media, an inverse fluorescence change is observed after the ATP-concentration jump, which is likely to be associated with the transition $E_2(K_2) \rightarrow K_2 \cdot E_1$. In contrast to the large electrical signals observed in the presence of Na^+, no transient current could be detected when the medium contained only K^+. This finding indicates that the deocclusion step ($E_2(K_2) \rightarrow K_2 \cdot E_1$) is electrically silent. The observation that Na^+ translocation is electrogenic, whereas K^+ translocation is electrically silent, can be explained assuming that the alkali-ion binding-site of the protein bears a charge of $-2e_0$, corresponding to a net charge of $+e_0$ in the sodium-loaded form and a net charge of zero in the potassium-loaded form. The conclusions which we have drawn from these experiments may be compared with the results of previous studies. From voltage-jump current-relaxation experiments with cardiac cells, Nakao and Gadsby (1986) proposed that translocation of Na^+ is a major charge-carrying step in the pumping cycle of the Na,K-ATPase. Similar experiments gave evidence that translocation of K^+ is an electrically silent process (Bahinski, Nakao & Gadsby, 1988). Essentially the same conclusion has been drawn by Goldshlegger et al. (1987) from the finding that in reconstituted vesicles potassium-potassium exchange is voltage insensitive. These observations basically agree with the results of the present study.

REFERENCES

Apell, H.-J., Borlinghaus, R., and Läuger, P. (1987). Fast charge-translocations associated with partial reactions of the Na,K-pump. II. Microscopic analysis of transient currents. J. Membrane Biol. **97**: 179-191

Apell, H.-J. (1989). Electrogenic properties of the Na,K-pump. J. Membr. Biol. (in press)

Bahinski, A., Nakao, M., and Gadsby, D.C. (1988). Potassium translocation by the Na/K pump is voltage insensitive. Proc. Natl. Acad. Sci. USA **85**: 3412-3416

Borlinghaus, R., Apell, H.-J., and Läuger, P. (1987). Fast charge-translocations associated with partial reactions of the Na,K-pump. I. Current and voltage transients after photochemical release of ATP. J. Membrane Biol. **97**: 161-178

Cantley, L.C., Carili, C.T., Smith, R.L., and Perlman, D. (1984). Conformational changes of Na,K-ATPase necessary for transport. Curr. Top. Membr. Transp. **19**: 315-322

Deguchi, N., Jørgensen, P.L., and Maunsbach, A.B. (1977). Ultrastructure of the sodium pump. Comparison of thin sectioning, negative staining, and freeze-fracture of purified, membrane-bound (Na^+,K^+)-ATPase. J. Cell Biol. **75**: 619-634

Fendler, K., Grell, E., Haubs, M., and Bamberg, E. (1985). Pump currents generated by the purified Na^+,K^+-ATPase from kidney on black lipid membranes. EMBO J. **4**: 3079-3085

Glynn, I.M., and Karlish, S.J.D. (1976). ATP hydrolysis associated with an uncoupled sodium flux through the sodium pump: Evidence for allosteric effects of intracellular ATP and extracellular sodium. J. Physiol. (London) **256**: 465-496

Glynn, I.M. (1985). In: The Enzymes of Biological Membranes, 2nd ed., Vol. 3, (A.N. Martonosi, ed.), Plenum, New York, pp. 35-114

Goldshlegger, R., Karlish, S.J.D., Rephaeli, A., and Stein, W.D. (1987). The effect of membrane potential on the mammalian sodium-potassium pump reconstituted into phospholipid vesicles. J. Physiol. (London) **387**: 331-355

Jørgensen, P.L., and Andersen, J.P. (1988). Structural basis for E_1-E_2 conformational transitions in Na,K-pump and Ca-pump proteins. J. Membrane Biol. **103**: 95-120

Kapakos, J.G., and Steinberg, M. (1986). 5-Iodoacetamidofluorescein-labeled (Na,K)-ATPase. Steady-state fluorescence during turnover. J. Biol. Chem. **261**: 2090-2096

Karlish, S.J.D., Yates, D.W., and Glynn, I.M. (1978). Elementary steps of the (Na^++K^+)-ATPase mechanism, studied with formycin nucleotides. Biochim. Biophys. Acta **525**: 230-251

Karlish, S.J.D. (1980). Characterization of conformational changes in (Na,K)-ATPase labeled with fluorescein at the active site. J. Bioenerg. Biomembr. **12**: 111-135

Läuger, P., and Apell, H.-J. (1988). Transient behaviour of the Na^+/K^+-pump: microscopic analysis of nonstationary ion-translocation. Biochim. Biophys. Acta **944**: 451-464

McCray, J.A., Herbette, L., Kihara, T., and Trentham, D.R. (1980). A new approach to time-resolved studies of ATP-requiring biological systems: Laserflash photolysis of caged ATP. Proc. Natl. Acad. Sci. USA **77**: 7237-7241

Nakao, M., and Gadsby, D.C. (1986). Voltage dependence of Na translocation by the Na/K pump. Nature (London) **323**: 628-630

Steinberg, M., and Karlish, S.J.D. (1989). Studies on conformational changes in Na,K-ATPase labeled with 5-iodoacetamidofluorescein. J. Biol. Chem. **264**: 2726-2734

Stürmer, W., Apell, H.-J., Wuddel, I., and Läuger, P. (1989). Conformational transitions and charge translocation by the Na,K pump: comparison of optical and electrical transients elicited by ATP-concentration jumps. J. Membr. Biol. (in press)

THEORETICAL AND EXPERIMENTAL ASPECTS OF THE SPECIFIC ASSOCIATION OF MONOVALENT IONS WITH LIPID BILAYER MEMBRANES

GREGOR CEVC

Medizinische Biophysik—Forschungslaboratorium, Urologische Klinik und Poliklinik, Klinik r.d.I., Technische Universität München, Ismaningerstr. 22 D-8000 München 80.

SUMMARY: The specificity of ion-surface association is likely to be determined by water mediated interactions, as can be seen from nonlocal electrostatic models, and experimental data, for lipid bilayer membranes. The strength of association between monovalent ions and charged polar surfaces decreases with ionic radius, owing to the more efficient dielectric screening of the electrostatic potential at greater separations. For non-charged surfaces the trend is reversed and the selectivity is governed by the free energy of ion dehydration.

INTRODUCTION

Surfaces of biomolecules and their aggregates usually carry net electric charges; such surfaces, moreover, typically contain numerous polar residues and, consequently, are extensively hydrated. These features play a rôle in, and depend upon, ion-surface interactions.

Long-range, coulombic electrostatic fields, which stem from the net surface charges, cause gathering of counterions of the opposite sign to that of the surface charges, typically cations, and dilution of ions of the same charge, coions, in the vicinity of the structural charges (Fig. 1A). Short-range interfacial fields, arising from the surface local excess charges located on the polar residues, to a large extent govern the molecular conformation, perturb the water structure in the interfacial region—because they are also involved in the hydrogen bonding with the water molecules—(Fig. 1B), and thus, in turn, influence the ion distribution.

All biological (macro)molecules are therefore at least partly embedded in diffuse double layers. The water structure, the ion concentration, and in many cases also the distribution of the surface associated charges in such layers all vary strongly with the separation from the interface. Furthermore, for individual ion types the distribution profiles are not the same: variable effective forces between ions and the macromolecular surface give rise to an ion-specific distribution and hence ion-specific effects. Especially for the relatively complex biological systems such specificity is crucial, as it often guarantees an optimal macromolecular structure and function.

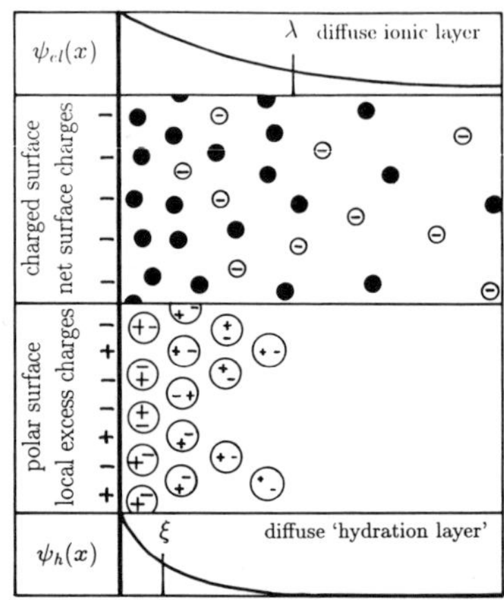

Figure 1: Schematic representation of the ion distribution near a charged surface (upper) and of the water distribution near a hydrophilic surface (lower).

The most familiar ion-surface interaction is ion binding. For all ions, but particularly for divalent and polyvalent ones, the *selectivity of binding* is believed to originate from quantum-mechanical interactions between the charged entity and a binding site (Gresh, 1980; Peinel et al., 1983). However, interactions between the resulting complexes and the water molecules often play a significant rôle as well (Pullman et al., 1977; 1978). To check experimentally whether or not for a given ion the specific binding has occurred, some system property—for example, the temperature of the transition between two lipid phases, the molecular packing density, the ion driven self-aggregation, etc—is plotted as a function of the bulk water activity; such a choice of the independent variable compensates for trivial non-interfacial effects (Cevc, 1988). If the resulting functional dependence deviates from a straight line, the non-linearity of the plot is indicative of ion binding (Fig. 2, the curved lines).

Experimentally observed linear deviations between the effects of various ions are a manifestation of the *selectivity of association*. This originates from the dissimilar probability for different ions, especially monovalent ones, to approach the macromolecular surface. This variability thus reflects disparities in the indirect ion-surface interactions and, consequently, is highly sensitive to the ionic double layer properties and to the overlap of the surface and ion hydration-layers. Experimentally, the specificity of association shows up in different slopes of the linear 'effect versus the bulk water activity' plots (Fig. 2, straight lines).

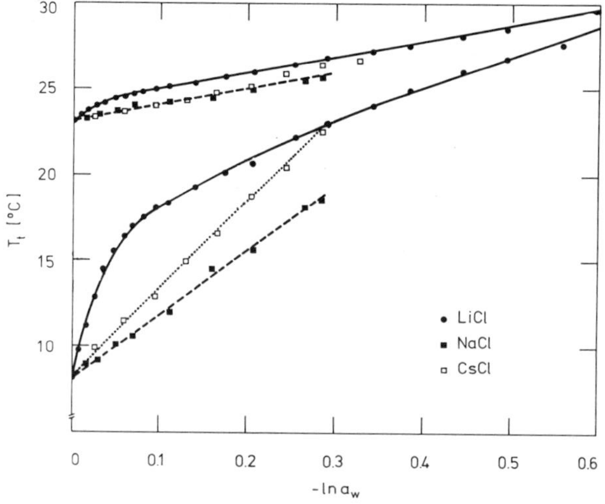

Figure 2: Effect of variable bulk salt concentration, here expressed in terms of the water activity coefficient a_w, on the main transition (upper group of curves) and cooling-scan pretransition temperature (lower group of curves) in dimyristoylphosphatidylcholine multibilayers. A non-linear functional dependence is diagnostic for lithium binding. The linearity of lines for sodium and cesium proves the absence of binding, the different slopes being indicative of differences in surface-ion association.

EXPERIMENTAL

All data presented in this work stem from measurements of different transition temperatures of > 99.9 % pure lipid bilayers, in particular, of the difference between the pretransition and the chain-melting phase transition temperature. They are based on using established techniques, most frequently on observing the temperature dependence of light scattering. The charged bilayer membranes were prepared by suspending 1,2-dimyristoylphosphatidyl-*sn*-glycero-3-phospho-DL-glycerol in 50 mm pH 7.5 buffer containing various amounts and types of salt, or sugars, for comparison. Uncharged membranes were made in a similar manner except in that the lipid was 1,2-dimyristoylphosphatidyl-*sn*-glycero-3-phosphocholine. Chloride was chosen as a co-ion because we have found that this minimizes any anion-specific effects. Compensation for bulk effects was achieved by comparing the data for electrolyte and sugar solutions. Focusing on the lipid pretransition, which is a low-enthalpy phase change, maximized the experimental sensitivity (see Fig. 2). This furthermore ensured that principally the interfacial ion concentration was being probed, owing to the specific origin of the lipid pretransition.

For the calculations simple programmes in Basic or Math-Cad internal language were run on a PC computer; this also included parameter optimization.

THEORETICAL ANALYSIS

Appropriate theoretical bases for assessing the specificity of ion *binding* are quantum-mechanics (Pullman et al., 1978; Gresh, 1980) and computer simulations (Peinel et al., 1983). To explain the specificity of indirect *association* between an ion and a surface, however, it normally suffices to use a simplified version of one of the basic equations of statistical mechanics together with an advanced electrical double layer model.

The required equation from the statistical theory of liquids relates the ion distribution profile to the bulk ion concentration and to the mean-potential of the ion-surface interactions, via a Boltzmann factor,

$$c_i(x) = c_{i,bulk} \exp[e\psi_{tot}(x)/kT] \tag{1}$$

The desired double layer model should allow for, in order of decreasing importance: 1. hydration phenomena, 2. ionic polarizability, 3. steric size-effects and 4. image-charge effects. The latter two contributions cause the interfacial ion concentration to decrease with ionic radius. The former two lead to a size-dependent concentration increase for a charged surface; for non-ionic surfaces, however, with increasing ion radius the hydration effects become smaller. In most systems the latter two (and often the last three) effects are relatively small. To clarify the mechanism of ion selectivity in the absence of direct binding, one should therefore concentrate on the first one or two listed effects, preferably unified in a suitably general model of hydration.

In electrostatic models, it is customary to consider the dielectric constant, ϵ, as the only solvent and hydration dependent parameter. Indeed, by suitably describing the relative solvent permittivity and its various functional dependencies, $\epsilon(var)$, one can devise a *phenomenological* description for all four major electrostatic solvent effects:

- Non-local response (direct water binding), $\epsilon(\mathbf{k})$

- Ion-field effects (ion-dependent interfacial polarization), $\epsilon(c)$

- Dielectric saturation ('coulombic' hydration, electrostatic solvent polarization), $\epsilon(\mathbf{E})$

- Dielectric discontinuity (image effects), $\epsilon(\mathbf{r})$

Of these four, the first effect is probably the most important. That is to say that any electrostatic description of the mechanism of selective ion-surface interactions must at the very least take into account the consequences of the interfacial hydration. This is mainly because of non-local solvent response to an applied field. [1]

Interfacial hydration can be described relatively easily by extending a non-local electrostatic model initially introduced for studying bulk electrolytes (Dogonadze & Kornyshev, 1972; Dogonadze et al., 1973; Kornyshev & Vorotyntsev, 1980) and later adopted for interfacial (Kornyshev & Vorotyntsev, 1981; Vorotynsev & Kornyshev, 1984; Marčelja, 1977; Gruen & Marčelja, 1983) and membrane studies (Cevc, 1985; Cevc & Marsh, 1987; Cevc, 1989a; Cevc, 1989b). Its basic philosophy is the following.

In classical electrostatics, valid for homogeneous dielectrics and structureless solvents, an external field of a given form creates a response of similar shape. In dielectrically non-homogeneous media, such as aqueous electrolytes, the same applied field produces a response which is skewed, due to the correlations and coupling between the solvent molecules. In other words: the response to an applied field is non-local and its range is proportional to the space in which the water 'structure fluctuations' are correlated.

Polar residues on hydrophilic (macro)molecular or membrane surfaces, or net surface charges therefore have an impact on the electrolyte solution which is not limited to the field's origin but rather which extends over a finite separation of the order of the solvent correlation length, ξ, and the electrostatic decay length λ, respectively. The spatial profile of the electrostatic potential near a polar uncharged surface reflects this clearly (cf. Fig. 3, lower). From ordinary electrostatic theory, such as Gouy-Chapman, one finds this potential to be zero everywhere. However, nonlocal electrostatics suggests that the potential close to the surface should deviate from zero in every case, and that in electrolytes it should change sign at a distance somewhat greater than the 'solvent structure decay length', ξ. Surface inhomogeneity may play a rôle but does not affect this conclusion qualitatively.

The non-coulombic 'potential of zero net charge' as well as the interfacial hydration both increase with the surface polarity. The latter, in the first approximation, can be taken to be proportional to the surface density of the local excess charge, σ_p. This is shown by the nonlocal electrostatic result for the polarity-induced potential

$$\psi(x, \sigma_p, \sigma_{el} = 0) = (\sigma_p \xi / \epsilon_0 \epsilon_\infty)(1 - \epsilon_\infty / \epsilon)[(\xi / \lambda) \exp(-x/\lambda) - \exp(-x/\xi)] \tag{2}$$

This result also indicates that increasing the bulk salt concentration diminishes this

---

[1]Such conclusion is reached for any reasonable solvent model. Qualitative implications of the theoretical analysis for a mixture of dipolar hard spheres and hard-sphere ions near an interface (Carnie & Chan, 1982; Lee & Ladanyi, 1987), for example, are similar to those obtained from a continuum dielectric theory assuming an inner interfacial layer with a dielectric constant much smaller than the bulk value (Levine et al., 1969), or from the nonlocal electrostatic approach used in this contribution.

198

potential, since the Debye screening length, $\lambda \sim 0.304c^{-\frac{1}{2}}$ nm (for a 1:1 electrolyte at 25 °C) then decreases. The decrease is less than for the standard coulombic electrostatic potential, however, due to the limited capacity of ions to screen the surface polar residues and their atomic charges. In this and in the following equations ϵ_0, $\epsilon \simeq 78$, and $\epsilon_\infty \simeq 2.5$ are the permittivity of free space, the static, and the high frequency dielectric constant of the solvent, respectively.

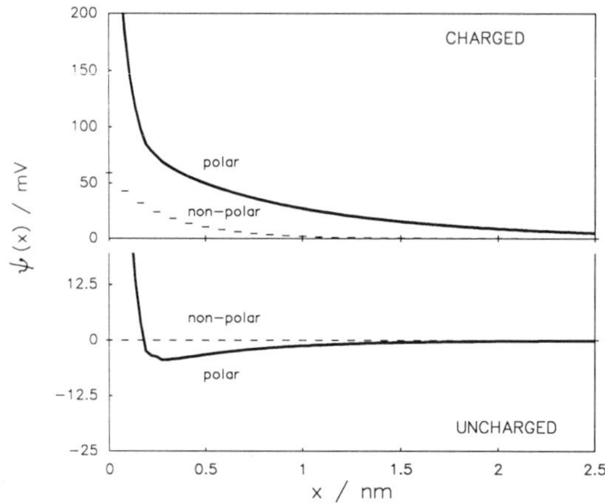

Figure 3: Electrostatic potential profile near a charged (upper) and uncharged (lower) surface in the standard approximation (dashed lines) and in the nonlocal electrostatic model which accounts for the surface polarity and solvent structure (full curves).

The coulombic electrostatic potential which arises from the net surface charges is always enhanced by the interfacial hydration (Fig. 3, upper part). This is caused by the diminished capability of the bound water molecules to dielectrically screen the net surface charges. To gauge the magnitude of the potential increase one can again use nonlocal electrostatics. In the linear approximation this yields for the electrostatic potential near a surface with a net surface charge density, σ_{el}, bathed by a dilute electrolyte solution:

$$\psi(x, \sigma_p = 0, \sigma_{el}) = (\sigma_{el}\lambda/\epsilon\epsilon_0)[\exp(-x/\lambda) + (\xi/\lambda)(\epsilon/\epsilon_\infty - 1)\exp(-x/\xi)] \tag{3}$$

The ion distribution profile, given by eq. 1, can be estimated simply by taking the total electrostatic potential of a charged polar surface (ψ_{tot}) to be the sum of the coulombic and polarity-dependent terms given by eqs. 2 and 3, respectively. Such approximation allows for the ion-size effects only implicitly. To obtain more accurate results the size of the test ion, at least, should be directly accounted for (Cevc, 1989b). Effects of linearization must also not be forgotten (Cevc & Marsh, 1987) but these only affect the final results quantitatively.

In other words, for a charged particle such as an ion which approaches the charged surface, the interphacial region between the (supra)molecular interior and the solution *appears* as if it was a region of reduced polarity and low dielectric constant with an excessively high electrostatic potential. In reality, of course, the interfacial dielectric constant is an ill-defined quantity, changing appreciably over separations comparable to the molecular dimensions of most of the system components (Vorotyntsev & Kornyshev, 1978). Moreover, this quantity is apt to change with all of the system parameters: with the interfacial charge density or polarity, with the interfacial separation and depth, with the molecular conformation, etc.

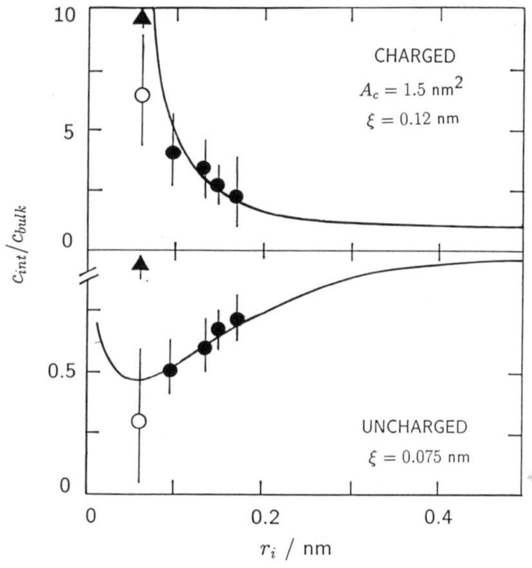

Figure 4: Experimental (symbols) and theoretical dependence of the relative ion concentration near the lipid bilayer surface as a function of ion radius, r_i. Curves were calculated by using a non-linear version of eqs. 2 and 3. Experimental values stem from the pretransition temperature measurements. (From Cevc, 1989b)

The strength of ion-surface association for charged macromolecules or their aggregates is seen, from a combination of eqs. 1, 2, and 3, to increase with decreasing ion radius, simply due to the fact that the smaller ions can climb higher up the electrostatic potential profile curve which also depends on the surface polarity (compare upper parts of Figs. 3 and 4). Qualitatively similar behaviour is expected for very small ions approaching a polar but non-ionic surface since such ions experience the local, hydrationally poorly screened part of the atomic excess charge potential (Figs. 3 and 4, lower parts); this holds true, however, only as long as the ionic radius does not exceed some critical size (in Fig. 4: 0.05 nm).

For larger ions the inner part of the hydrational potential profile is inaccessible. In this case, the energetic cost of dehydration, ie, the work spent during ion transfer from the bulk region, with a high polarity, into the less polar interfacial space, reverses the sequence of selectivity. Large ions interacting with a polar but uncharged surface are thus repelled from the interfacial region. The larger they are, the less they are repelled, owing to the decrease of the free energy of ion hydration with ionic radius. The curves and data points in the lower part of Fig. 4 vindicate this conclusion.

The data presented in this contribution all pertain to lipid bilayers. However, the underlaying concept of indirect ion-specific interactions is of general validity. It is therefore applicable to studies of any hydrophilic surface, such as the water-exposed surfaces of proteins, nucleic acids, etc.

Acknowledgement: This study has been supported by Deutsche Forschungsgemeinschaft under grants SFB/C8 and Ce 19/1-1.

REFERENCES

Carnie, S. L. & Chan, D. Y. C. (1982) J. Chem. Soc. Faraday II 78, 695-722.

Cevc, G. (1985) Chem. Scripta 25, 97-106.

Cevc, G. & Marsh, D. (1987) Phospholipid Bilayers. Physical Principles and Models, Wiley-Interscience, New York.

Cevc, G. (1988) Ber. Bunsen Ges. 92, 953-961.

Cevc, G. (1989a) J. de Physique 50, 1117-1134.

Cevc, G. (1989b) Submitted for publication.

Dogonadze, R. R. & Kornyshev, A. A. (1972) J. Chem. Soc. Faraday Trans. 2 70, 1121-1132.

Dogonadze, R. R., Kornyshev, A. A., & Kuznetsov, A. M. (1973) Theor. i Mat. Fiz. 15, 127-138.

Gresh, N. (1980) Biochim. Biophys. Acta 597, 345-357.

Gruen, D. W. R. & Marčelja, S. (1983) J. Chem. Soc. Faraday Trans. II 79, 225-242.

Kornyshev, A. A. & Vorotyntsev, M. A. (1980) Surf. Sci. 101, 23-48.

Kornyshev, A. A. & Vorotyntsev, M. A. (1981) Can. J. Chem. 59, 2031-2042.

Lee, P. H. & Ladanyi B. M. (1987) J. Chem. Phys. 87, 4093-4099.

Levine, S., Bell, g. M., Smith, A. L. (1969) J. Phys. Chem. 73, 3534-3545.

Marčelja, S. (1977) Croat. Chem. Acta 49, 347-358.

Peinel, G., Frischleder, H. & Binder, H. (1983) Chem. Phys. Lipids 33, 195-205.

Pullman, B., Gresh, N., Berthod, H. & Pullman, A. (1977) Theor. Chim. Acta 44, 151-163.

Pullman, A., Pullman, B. & Berthod, H. (1978) Theor. Chim. Acta 47, 175-192.

Vorotyntsev, M. A. & Kornyshev, A. A. (1978) Elektrokhimiya 15, 660-664.

Vorotynsev, M.A. & Kornyshev, A.A. (1984) Elektrokhimiya (transl) 20, 1-44.

THE ROLE OF ISOTHERMAL ENTHALPY PRODUCTION IN COUPLED SOLUTE TRANSLOCATION ACROSS THE MEMBRANE

Anjan Kr. Dasgupta

Department of Biochemistry and Biophysics, University of Kalyani, Kalyani 741 235, West Bengal, India

SUMMARY : Enthalpy change is shown to play a crucial in dictating the extent of coupling in the energy transduction and metabolically coupled solute translocation across biomembranes. At certain critical enthalpy values coupling would cease to exist. The chemical agents which induce critical enthalpy changes may thus be regarded as "uncouplers". The above conclusion follows from a general thermodynamic analysis which suggests that for heat exchanging nonequilibrium systems the entropy change, instead of chemical affinity, should be regarded as the macroscopic chemical force. As a corollary it follows that if the coupling reactions are stoichiometrically unbalanced, the structural ordering and disordering of the solvent molecules would constitute and integral part of the coupling process.

INTRODUCTION

Most of the metabolically coupled systems involve changes of enthalpy at stages away from equilibrium. The need of a thermodynamic theory that enables us to calculate the enthalpy change for a nonequilibrium process is thus obvious. The problem of isothermal heat flow has been treated in classical thermodynamic text books in terms of heat carried by solutes, and is described by Sorret coefficients (Groot and Mazur, 1962). One measures approximately the heat flux coupled to the diffusion flux. Such diffusion coupled heat flow may not be sufficient to describe processes like thermogenesis (Klingeniberg, 1982) which perhaps involves a metabolic control of the flow of heat. The explanation often offered to describe the thermogenesis is that the uncoupling of the oxidative phosphorylation is responsible for such regulated heat liberation. This commonsense explanation of heat flow associated with uncoupling may lead to some confusion. The argument that is taken for granted in correlating the heat generation with uncoupling is that the unused energy of the "driving" component of coupled reaction is converted to heat, at the uncoupled state. In the light of the chemiosmotic theory the heat generation associated with uncoupling will then be related to the dissipation of the proton gradient (Leninger, 1982; Lin & Klingenberg, 1987). Since the chemiosmotic explanation of uncoupling is the enhancement of protonic conductivity (Nicholls, 1982), using the analogy of a resistor carrying a current, it becomes difficult to appreciate how steady state heat flow would increase when the resistance decreases. In our view such confusion stems from the inadequacy of the existing theories to phenomenologically describe the enthalpy flow coupled to macroscopic forces and fluxes operative in a coupled reaction network.

GENERALIZATION OF HEAT BALANCE EQUATION

Let a given chemical system comprise of "r" coupled chemical reactions. The relation between the reaction affinity (A), molar free energy change (Δ G) and the chemical potential (μ_i) can be expressed as (Gray, 1970),

$$A_p = -\Delta G_p = - \sum_{j=1}^{n} \nu_{ip} \, \mu_j, \quad (p=1,2 \ldots r),$$

where, "p" is the suffix for the reaction, "j" is the suffix for the component and $\quad$ jp is the stoichiometric coefficient of the j'th component participating in the p'th reaction. To derive the local heat balance equation we consider,

$$d/dt \ (Q) = \int \dot{Q}v.dv, \qquad \ldots \ (1)$$

where, Q represents the rate of the total heat flow and Qv represents the volume density of the same. The volume contribution obviously arises due to enthalpy production associated with the coupling reactions. Thus,

$$\int_V \dot{Q}v.dv = \int_V H_p.d/dt \ (\xi p) \ dv = \int_V \Delta H_p.V_p.dv, \qquad \ldots \ (2)$$

where, $d/dt \ (\xi p) = V_p$, is the rate of change of reaction coordinate ξ p of the p'th chemical reaction, and Δ H$_p$ is the enthalpy difference of the products and the reactants of the p'th reaction. We can express the heat balance equation as;

$$\dot{Q} = \int_V d/dt \ (Qv).dv = \int_V (\delta/\delta t \ (Qv) + \nabla.Jq).dv, \qquad \ldots \ (3)$$

where, the divergent term arises due to net translational motion occurring across the boundary and $\delta Qv/\delta t$ represents the expdlicit rate of change of Qv with time. The heat balance equation can thus be expressed as;

$$\delta/\delta t \, (Qv) = -\nabla.Jq + \sum_{p=1}^{r} Vp.\Delta Hp \qquad \qquad \dots \quad (4)$$

The entropy and mass balance equations are respectively given by (Katchalsky and Curran, 1967);

$$\delta/\delta t \, (Sv) = -\nabla.Js + \sigma \qquad \qquad \dots \quad (5)$$

$$\delta/\delta t \, (Ci) = -\nabla.Ji + \sum_{p=1}^{r} \nu_{ip}.Vp, \qquad \qquad \dots \quad (6)$$

where, Ci is the concentration of the ith component, Ji is the diffusional flux of the i'th component σ is the internal entropy production, Sv is the volume density of entropy and Js is the entropy flux. The local Gibbs equation (Katchalsky and Curran, 1967) can be expressed as;

$$\delta/\delta t \, (Qv) = T.\delta/\delta t \, (Sv) + \sum_{i=1}^{n} \tilde{\mu}_i \, \delta/\delta t \, (Ci) \qquad \qquad \dots \quad (7a)$$

From equations (4-7);

$$-\nabla.Jq + \sum_{P=1}^{r} Vp.\Delta Hp = T.(\sigma-\nabla.Js) + \sum_{i=1}^{n} (\nu_{ip}.Vp - \nabla.J_i)\tilde{\mu}_i \qquad \dots \quad (7b)$$

Assuming, ΔSP to be the molar entropy change and ΔGp to be the molar free energy change of the p'th reaction we obtain;

$$\Delta Sp = \sum_{i=1}^{n} \nu_{ip}.Si, \qquad \qquad \dots \quad (8a)$$

$$\Delta Gp = \sum_{i=1}^{n} \nu_{ip} \tilde{\mu}_i, \qquad \qquad \dots \quad (8b)$$

where,

$$\Delta Gp = \Delta Hp - T.\Delta Sp, \qquad \qquad \dots \quad (8c)$$

Substituting equations (8a)-(8c) in equation (7b) and then comparing the divergent and nondivergent terms, we obtain;

$$Jq = T.Js - \sum_{i=1}^{n} \tilde{\mu}_i.Ji, \qquad \qquad \dots \quad (9a)$$

$$\sigma = -Jq.\nabla(1/T) + \sum_{i=1}^{n} Ji.\nabla(-\mu_i/T) + \sum_{p=1}^{r} Vp.\Delta Sp \qquad \dots \quad (9b)$$

An interesting divergence from the conventional dissipation equation emerges in the term representing the chemical reaction. The driving force for the reaction flux is given by the molar entropy change ΔS_p and not by the molar free energy change of the p'th reaction as it is conventionally believed (Gray, 1970; Shear, 1968).

SOLVENT INTERPRETATION OF THE ENTROPIC DRIVING FORCE

THE NEW PHENOMENOLOGICAL EQUATION FOR A CHEMICAL SYSTEM:

If we consider the diffusional force or, the temperature gradient to be absent, the dissipation is given by the expression $\sum_{p} Vp.\Delta Sp$. The phenomenological equation which follows is :

$$Vp = \sum_{p'} Lpp'\Delta Sp' \qquad \qquad \dots \quad (10)$$

Equation (10) obviously suggests that the kinetics of the system is dictated by the molar entropy changes. The argument that ΔGP provides the direction of the reaction however still remains valid. Expressing the Law of mass action for forward and reverse reaction velocities (Vp + and Vp-), the free energy change is given by;

$$\Delta Gp = -RT.\ln(Vp+/Vp-) \qquad \qquad \dots \quad (11)$$

Equation (11) suggests that as long as the reaction proceeds in the forward direction i.e., $Vp = (Vp+)-(Vp-) > 0$; ΔGP must have a negative sign.

SOLVENT FREE ENERGY CHANGE AND THE ENTROPIC DRIVING FORCE :

Consider the chemical system with unit volume. The expression for total free energy is given by;

$$G = \sum_i \tilde{\mu}_i . Ci \qquad \qquad \dots (12)$$

For a reaction comprising of n components with p reaction steps there is one component, namely the solvent, which deserves a special status as the solvent concentration Cw may not be an exclusive function of the reaction coordinates. The chemical potential for the ith component of a chemical system can be expressed in the ideal solution limit as (Glasstone, 1942);

$$\tilde{\mu}_i = \tilde{\mu}_i^o + \tilde{V}i.P + \tilde{S}i.T + RT.\ln f_i, \qquad \dots (13a)$$

where, the mole fraction f_i is given by;

$$f_i = Ci / \sum_{i=1}^{n} Ci \qquad \qquad \dots (13b)$$

An additional potential term needs to be added to equation (13b) if the component under consideration, is ionic. The rate of free energy change of the solvent is expressible as;

$$\dot{G}w = d/dt \, (Cw. \, \tilde{\mu}_w) = Cw.d/dt \, (\tilde{\mu}_w) \qquad \dots (14)$$

Since,

$$d/d \, \xi \, p \, (Ci) = \nu ip, \qquad \qquad \dots (15)$$

assuming that $\tilde{V}w \simeq 1/Cw$, and Sw (th molar entropy of the solvent) to be independent of the reaction coordinate $\xi \, p$, we obtain;

$$\dot{G}w = \sum_{p=1}^{r} Vp. \, \delta / \delta \xi \, p \, (Gw) = \sum_{p=1}^{r} Vp \Delta \, Gwp, \qquad \dots (16a)$$

where,

$$\Delta \, Gwp = \delta / \delta \xi \, p \, (Gw) = -RTf_w \Delta \, Np. \qquad \dots (16b)$$

206

In equation (16b) the quantity ΔNp represents the algebric sum of the stoichiometric coefficients of the p'th reaction. In other words,

$$\Delta Np = \sum_{i=1}^{n} \nu_{ip} \qquad \qquad \ldots \quad (17)$$

ΔNp will vanish only if there is a stoichiometric balance between the reactants and products. Such stoichiometric balance exists in unimolecular reactions or, group transfer reactions. If we ignore the fractional volume change of the reacting system we can assume the approximation;

$$\Delta v_p = \sum_{i=1}^{n} \nu_{ip} \, \tilde{v}_i \simeq 0 \qquad \qquad \ldots \quad (18)$$

Substituting equations (13a) and (13b) in equation (8b) we obtain,

$$\Delta Gp = (\Delta Gp^{O} + RT \sum_{i} \nu_{ip} \ln Ci) + T \Delta Sp - RTCw \, \Delta Np \qquad \ldots \quad (19a)$$

where, we have used the following equations.

$$\Delta Gp^{O} = \sum_{i=1}^{n} \nu_{ip} \, \mu_i^{o} \qquad \qquad \ldots \quad (19b)$$

$$\text{and,} \quad \sum_{i=1}^{n} Ci = Cw + \sum_{i \neq w}^{n} Ci \simeq Cw \qquad \ldots \quad (19c)$$

The approximation in equation (19c) is valid as long as the reacting system is dilute. The RHS of equation (19b) can be identified as the standard state free energy. Since the standard state is conventionally chosen, so as to obtain the expression for free energy change for pth reaction given by the first bracketted term of the RHS of equation (19a), we obtain the relation;

$$T \Delta Sp = RTCw \, \Delta Np. \qquad \qquad \ldots \quad (20)$$

If equation (19c) is valid f_w tends to unity in the lower dilution limit. Under such a condition we obtain the using equations (16b) and (20);

$$T \, \Delta Sp = -Cw \, \Delta \, Gwp. \qquad \ldots \quad (21)$$

Equation (21) expresses the important relation that the entropic driving force Δ Sp, can be equated to be molar free energy change of the solvent multiplied by the sign changed solvent concentration.

SOME PREDICTIONS OF THE NEW PHENOMENOLOGICAL DESCRIPTION :

An important thermodynamic result that follows, is, a group transfer reaction or a unimolecular reaction cannot proceed in an uncoupled state. This is a consequence of equation (20) according which if Δ Np = 0, the entropy change Δ Sp vanishes. A "lone" reaction cannot proceed in forward or reverse direction if the macroscopic force driving such a reaction vanishes. The result trivially follows from the simplification of equation (10) for a "lone" reaction :

$$Vp = Lp \, \Delta \, Sp = RLpCw \, \Delta \, Np \qquad \ldots \quad (22)$$

If Sp vanishes naturally Vp assumes a null value.

The second result that deserves mention follows from the argument following equations (11);

$$RT \, \ln \, (Vp+/Vp-) \, > \, 0 \, \text{ for } Vp+ > Vp-; \qquad \ldots \quad (23a)$$

Combining equations (11), (20) and (23a) we obtain;

$$\Delta Hp \, < \, RTCw \, \Delta \, Np. \qquad \ldots \quad (23b)$$

Alternatively, one can express (23b) as :

$$- \, \Delta \, Hp \, > \, Cw \, \Delta \, Gwp. \qquad \ldots \quad (23c)$$

Equation (23c) expresses that the minimum heat liberated from pth reaction is related to the solvent free energy change of the same.

ENERGY TRANSDUCING SYSTEMS

The energy transducing systems have been phenomenologically described by a number of earlier workers (Rottenberg, 1979; Vandam et al., 1980). In each case however the role of heat flow has been ignored and the authors considered, the conventional phenomenological equation in which the macroscopic force is equated with the chemical affinity or the free energy change. Using the Katchalsky-convention for discrete systems (Katchalsky and Curran, 1967) the dissipation equation can be described as :

$$\sigma = Jq. \, \Delta \, (1/T) + \sum_{i=1}^{n} \, Ji.\Delta \, (\tilde{\mu}i/T) + \sum_{p=1}^{r} \, Vp.\delta \, Sp. \qquad \ldots \quad (24)$$

In equation (24) we have assumed two difference symbols Δ and δ , representing respectively, the gradient along the membrane thickness, and along the reaction coordinate. Now the phenomenological equations for a general energy coupled solute translocation system can be derived from the dissipation equation (24).

$$\Delta \, (1/T) \, = \, Rq \, Jq + \sum_{i=1}^{n} \, Rqi \, Ji + \sum_{p=1}^{r} \, Rqp \, Vp, \qquad \ldots \quad (25a)$$

$$\Delta \, (\, \tilde{\mu}_{i}/T) \, = \, Rqi \, Jq + \sum_{i=1}^{n} \, Rij \, Jj + \sum_{p=1}^{r} \, Rip \, Vp, \qquad \ldots \quad (25b)$$

$$\delta Sp \, = \, Rqp \, Jq + \sum_{i=1}^{n} \, Rip \, Ji + \sum_{p'=1}^{r} \, Rpp' \, Vp' \qquad \ldots \quad (25c)$$

We first assume existence of a steady state flow of heat for which the explicit temporal dependence of the heat flow must vanish. In other words,

$$\delta/\delta \, t \, (Qv) \, = \, 0 \qquad \ldots \quad (26a)$$

Substituting equation (26a) in equation (4), we obtain after integrtation

$$\int_{s} Jq. \, ds \, = \, \int_{v} \sum_{p=1}^{r} \, Vp.\delta \, Hp. \, dv. \qquad \ldots \quad (26b)$$

Using the approximation that Jq is replaceable by its surface average and Vp & Δ Hp can be replaced by their respective volume averages, we obtain the equation;

$$Jq = x^{-1} \sum_{p=1}^{r} Vp.\delta Hp, \qquad \ldots \ (26c)$$

where, x is the surface to volume ratio. Substituting equation (26c) in equation (25c) we obtain;

$$\delta Sp = \sum_{p'=1}^{r} r_{pp'} \, Vp' + \sum_{i=1}^{n} Rip \, Ji, \qquad \ldots \ (27a)$$

where, $r_{pp'}$ is given by the equation;

$$r_{pp'} = x^{-1} Rqp \, \delta Hp' + Rpp' \qquad \ldots \ (27b)$$

Using the Onsager symmetry relation $Rpp' = Rp'p$, for the phenomenological equations (25a, 25b, 25c) and using the symmetry relation for the reduced phenomenological equation described by equation (27) we obtain the relation;

$$Rqp \, \delta Hp' = Rqp' \, \delta Hp \qquad \ldots \ (28)$$

Equation (28) implies that the phenomenological coefficient Rqp can be expressed as;

$$Rqp = a.\delta Hp, \qquad \ldots \ (29)$$

where, 'a' is a constant. Using equations (29) and (25c) we obtain the following reduced equation:

$$r_{pp'} = R_{pp'} + a.x^{-1}.\delta Hp.\delta Hp' \qquad \ldots \ (30)$$

Equation (30) expresses that the p'th and p'th reactions are coupled partially by the enthalpy changes of the respective reactions and the magnitude of such coupling is proportional to the product of the enthalpies of the respective reactions. In other words, an indirect coupling between two reactive pathways can be mediated by enthalpy exchange occurring at non-equilibrium state.

210

<u>OXIDATIVE PHOSPHORYLATION</u> :

The phenomenological equations described by equation (25a)-(25c), can be applied to describe the Oxidative Phosphorylation process (Rottenberg, 1979; Stucki, 1980). The component index can be either "O" or, "P" for the Oxidative and Phosphorylative process respectively. The Proton translocation is symbolized by "H" which replaces the index "i" in equations (25a)-(25c). The reduced coefficient "r" is written instead of "R" whenever a contraction is implied by equation (26c). We obtain the following equations :

$$\Delta \, (1/T) = R_{qH} \, J_H + r_{qP} \, V_p + R_{qO} \, V_o \qquad \qquad \ldots \quad (31a)$$

$$(\Delta \tilde{\mu}_H) = R_H \, J_H + r_{HP} \, V_P + r_{HO} \, V_O \qquad \qquad \ldots \quad (31b)$$

$$\delta S_o = R_{OH} J_H + r_{OP} \, V_P + r_{OO} \, V_O \qquad \qquad \ldots \quad (31c)$$

$$\delta S_P = R_p H \, J_H + r_{PP} \, V_P + r_{PO} \, V_O \qquad \qquad \ldots \quad (31d)$$

Let us now consider the relations between the original and reduced phenomenological coefficients. Using equation (30) the coefficient r_{OP} is given by

$$r_{OP} = R_{OP} + ax^{-1} \, \delta H_O . \, \delta H_P \qquad \qquad \ldots \quad (32)$$

The other cross coupling coefficients are given by [using equation (29)];

$$r_{qP} = R_{qP} + R_q \, x^{-1} \, \delta H_P = a.\delta H_P + R_q \, x^{-1} \, \delta H_P \qquad \ldots \quad (33a)$$

$$r_{qO} = R_{qO} + R_q \, x^{-1} \, \delta H_o = a \, \delta H_o + R_q \, x^{-1} \, \delta H_o \qquad \ldots \quad (33b)$$

$$r_{HP} = R_{HP} + R_{qH} \, x^{-1} \, \delta H_P \qquad \qquad \ldots \quad (33c)$$

$$r_{HO} = R_{HO} + R_{qH} \, x^{-1} \, \delta H_o \qquad \qquad \ldots \quad (33d)$$

If we assume furthermore that $\Delta (1/T) = 0$, i.e. no temperature gradient exist for arbitary values of J_H, V_P or V_O then the coefficients R_{qH}, r_{qP}, r_{qO} vanish. The Onsager symmetry imposes no further restriction to the coefficient and one obtains the simplified relations;

$$a = - R_q x^{-1}, \qquad \dots \ (34a)$$

$$r_{OP} = r_{PO} = R_{OP} - x^{-1} R_q . \delta H_p . \delta H_0 \qquad \dots \ (34b)$$

$$r_{HP} = R_{HP} \qquad \dots \ (34c)$$

$$r_{HO} = R_{HO} \qquad \dots \ (34d)$$

R_q being the diagonal term of the phenomenlogical matrix is postitive, and a close examination reveals that it is reciprocally related to the thermal conductivity of the biomembrane. If "q" represents the degree of coupling the relations which follow are $q_{OH} = -r_{OH}/(r_{OO} R_H)^{1/2}$, $q_{pH} = - r_{pH} / (r_{PP} R_H)^{1/2}$ and $q_{OP} = - r_{OP}/(r_{OO}.r_{PP})^{1/2}$. If Mitchell's hypothesis, that no direct coupling between the Oxidative and Phosphorylative process exists, is assumed, following Rottenberg (Rottenberg, 1979), we can employ the equation :

$$q_{OP} = - q_{OH} . q_{PH} \qquad \dots \ (35)$$

$$R_{OP} = x^{-1} R_q \delta H_O \delta H_P + R_H^{-1} r_{OH} r_{PH} \qquad \dots \ (36)$$

Equation (36) reveals the interesting fact that the net observed coupling between the Electron transport and ATP -synthesis pathways comprises two major components, one depending on the thermal conductivity, and the other depending on the Proton permeability of the membrane. If the protonic resistance is high ($R_H \to \infty$) the former dominates. By imposing various degrees of thermal insulation of the membrane (by say, addition of external agents)one can induce different degrees of coupling or uncoupling, and this acts independently of the Proton circuit postulated by Mitchell.

REFERENCES

Glasstone, (1942) In : Introduction to electrochemistry, Van Nostrand, N.Y.

Gray, B.F. (1970) Trans Faraday Soc. 66, 363.

Groot, S.R., and Mazur, P. (1962) In : Non-Equilibrium Thermodynamics, North Holland, Publishing Company, pp. 273-284.

Katchalsky, A. and Curran, P.F. (1967) In : Non-equilibrium thermodynamics in biophysics Harvard University Press.

Klingenberg, M. (1987) Bioenergetics : Structure and Function of Energy Transducing Systems (T. Ozawa and S. Papa Eds). Japan Science, Soc. Press, Tokyo/Springer-Verlag, Berlin, pp. 3-17.

Leninger, A.L., (1982) In : Principles of Biochemistry. Worth Publishers, New York.

Lin, C.S. and Klingenberg, M. Biochemistry, 21, 2950.

Nicholls, D.G. (1982) In : Bioenergetics, Academic Press, London.

Rottenberg, H. (1979), Biochemica et Biophysica Acta, 549, 225.

Shear, D. (1968) J. Chem. Physics. 48, 4144.

Stucki, J.N. (1980), Eur. J. Biochem. 109, 269.

Vandam, K., Westerhoff, H.V., Krab, K., Meer, V.D., and Arents, J.C. (1980) Biochemica et Biophysica Acta, 591, 240.

INTRACELLULAR SPERMINE MODIFIES NEURONAL ELECTRICAL ACTIVITY

H. Drouin[+]) and A. Hermann[*])

[+])Institut für Physiologie, Medizinische Universität zu Lübeck,
  Ratzeburger Allee 160, D-2400 Lübeck 1, F.R.G.
[*])Institut für Zoologie, Abteilung Physiologie, Universität
  Salzburg, Hellbrunner Strasse 34, A-5020 Salzburg, Austria

SUMMARY: The polyamine, spermine, if ionophoretically injected into
identified molluscan neurons, reduces the amplitude of action
potentials and alters their wave form. Spermine cations block both,
the calcium inward current and the delayed potassium outward current
in a dose dependent manner. The block of the ionic currents by
spermine cations can be described as a voltage dependent reaction.

INTRODUCTION

Spermine, with the formula: $NH_3^+(CH_2)_3NH_2^+(CH_2)_4NH_2^+(CH_2)_3NH_3^+$, is
classified as a polyamine which bears four positive charges at
physiological pH values. Polyamines, such as putrescine, spermidine or spermine, interact with anionic components of natural
membranes such as membrane-bound enzymes; they affect transport
of ions and metabolites, influence cellular calcium homeostasis,
interfere with polyphosphoinositide metabolism and protein kinase
C reactions, bind to phospholipids and act on membrane fusion
(for review, see Schuber, 1989).

Polyamines have been identified in excitable cells where they
play a role in nerve growth, nerve regeneration and survival of
nerve cells (Slotkin & Bartolome, 1986; Gilad & Gilad, 1988).
After electrical stimulation of nervous tissue, the biosynthesis
of polyamines is augmented (Gould & Cottrell, 1974; Pajunen et
al., 1978) and in studies with molluscan giant neurons it was
found that ionophoretically injected tritium-labeled putrescine

was rapidly metabolized to spermidine and spermine (Kremzner & Ambron, 1982).

There is, however, little information concerning the action of spermine cations on electrical activity of nerve cells. The purpose of this investigation was to study the effects of spermine on action potentials and on ionic currents in more detail (Drouin & Hermann, 1984).

MATERIALS AND METHODS

Identified nerve cells in the abdominal ganglion of Aplysia californica were used for electrophysiological investigation (Kandel, 1976; Drouin, 1983). A nerve cell was impaled with up to four microelectrodes for recording membrane potential, passing current, and ionophoretic injection of spermine cations and/or calcium or EGTA (ethylene-glycol tetraacetic acid). Voltage clamp and current measuring circuitry were identical to those described previously (Hermann & Gorman, 1981).

RESULTS AND DISCUSSION

Effects of spermine: Fig.1 summarizes the effects of intracellular spermine cations on spontaneous electrical activity of the bursting pacemaker cell R-15. The left part of Fig.1 shows bursting activity and the right part indicates superimposed action potentials which correspond to the bursts marked by the inverted triangles. The top records (C = control) show pacemaker activity under normal extra- and intracellular ionic conditions. The lower part 1 of Fig.1 represents records, taken after an amount of 3.5 mM of spermine was injected ionophoretically. The horizontal line through the traces marks zero membrane potential. Ionophoretic injection of spermine cations into nerve cells reduces amplitudes of action potentials and removes the hump in their falling phase (Gorman &Hermann,1982) as seen in the upper right part of Fig.1C.

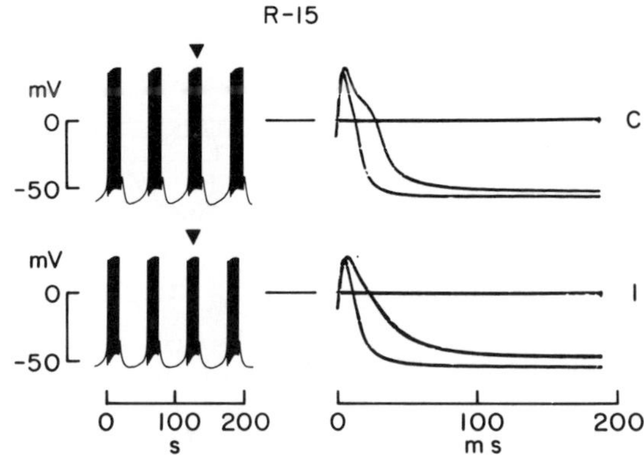

Figure 1. Effects of intracellular spermine cations
on bursting activity of cell R-15

These results indicate that internal spermine affects various
membrane currents underlying action potentials. Two major currents which determine the shape of action potentials are the
calcium inward current and the delayed potassium outward current.
Therefore, we studied these currents after their pharmacological
separation from other superimposing membrane currents.

The lower part of Fig.2 shows the relationship between the
isolated delayed potassium outward currents, I_K, and voltage, V,
under control conditions (filled symbols in C) and after ionophoretic injection of various amounts of spermine cations (open symbols) at 200 nA for 4 min (1), 6 min (2) and 8 min (3). The upper
part of Fig.2 shows the relationship between normalized potassium
currents and membrane voltage after injection of various amounts
of spermine cations. The drawn lines are calculated according to
Eq.2 and the parameters specified below.

The lower part of Fig.3 shows the relationship between the
maximum calcium inward currents, I_{Ca}, and voltage, V, under control conditions (filled symbols in C) and after ionophoretic injection of spermine (open symbols), at 200 nA for 4 min at (1) and
2 min at (2) to (5). The upper part of Fig.3 shows the relationship between normalized calcium currents and voltage with increasing concentration of intracellular spermine cations. The
lines through the experimental points are calculated according to

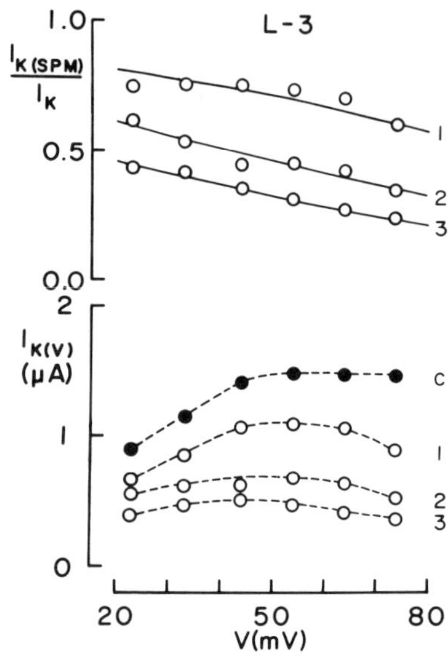

Figure 2.
Action of intracellular
spermine cations on delayed
potassium outward currents
of cell L-3

Figure 3.
Action of intracellular
spermine cations on maximum
calcium inward currents
of cell L-6

Eq.2 and the parameters specified below.

The results indicate that intracellular spermine blocks the delayed potassium outward current as well as the calcium inward current in a dose and voltage dependent manner.

Reaction model: The block of ionic membrane currents can be described by a voltage dependent reaction (Drouin, 1983), where n spermine cations, SPM, bind to one channel, R, and form a blocking complex, SPM_nR,

$$n \; SPM + R \; = \; SPM_nR \; . \tag{1}$$

The total concentration of injected spermine, $[SPM]_{inj}$, is assumed to be equal to the sum of concentrations of free and bound spermine. The total concentration of reaction sites is thought to be equal to the sum of concentrations of unoccupied and occupied reaction sites. Further it is postulated that the fraction of unoccupied reaction sites is equal to the fraction of unblocked channels. These considerations result in the equation,

$$I(SPM)/I = \{1 + ([SPM]_{inj} - n\,[SPM_nR])^n K^0 exp(V/V_B)\}^{-1}, \qquad (2)$$

which is taken for the analysis of normalized currents as function of intracellular concentration of spermine cations and the total voltage difference, V, across the membrane. Eq.2 is determined by four parameters: the binding constant, K^0, at zero voltage, the concentration of bound spermine, $[SPM_nR]$, the coefficient, n, that is allowed to have nonintegral values, and the experimental voltage parameter, V_B.

The effective valence, $z\delta$, of spermine cations which determines the voltage dependence of the block can be calculated according to the relation,

$$z\delta = RT/FVB . \qquad (3)$$

z is the valence of spermine cations; δ is the fraction of the total voltage difference, V, seen at the blocking site; F is the Faraday constant; R is the gas constant and T is the absolute temperature.

<u>Potassium currents</u>: The concentrations of injected spermine cations in the nerve cells were estimated from optical measurements of cell diameters and use of the equation for microelectrode ionophoresis (Drouin, 1984). The geometry of cell L-3 was approximated by a prolate spheroid, and gave a resulting cell volume of 30 nl. The ionophoretic current of 200 nA, therefore, generated an average concentration of 1.44 mM at 4 min, 3.6 mM at 10 min and 6.5 mM at 18 min. The upper part of Fig.2 shows how experimental values of normalized potassium outward currents and voltage under the influence of injected spermine cations can be described by Eq.2. The parameters as found from fitting Eq.2 to the experi-

mental data are: $K_b=K^0=140/M$; $n=1$; $[SPM-R]=0.26$ mM; $z\delta=0.45$.

The mechanism of action of spermine as a charged molecule can be explained by simple obstruction of the channel and voltage dependent promotion of the molecule to its binding sites. Following Miller's (1982) reasoning, we suppose that this compound blocks in a conformation with its four positive charges, z_1, z_2, z_3 and z_4, separated in space inside the channel, such that each charge experiences a different fraction, δ_1, δ_2, δ_3 and δ_4, of the electric voltage difference. Each charge will make an independent contribution to the voltage dependence of the binding energy in proportion to its δ, so that the observed effective valence will be, $z\delta=z_1\delta_1+z_2\delta_2+z_3\delta_3+z_4\delta_4$. For the spermine cation, $z_1=z_2=z_3=z_4=1$ and, therefore, $z\delta=\delta_1+\delta_2+\delta_3+\delta_4$. In the case that $\delta=\delta_1=\delta_2=\delta_3=\delta_4=0.5$ and the four charges close together, the effective valence should be high, approaching the value 2, as the charges superimpose upon each other. However, as the separation of the charges increases, the effective valence should drop as the second, third and fourth chrges are left behind less deeply inside the channel. This decrease in effective valence should occur until the fourth, third and second charges are left completely outside the applied electric field ($\delta_4=\delta_3=\delta_2=0$), though perhaps not outside the channel protein. However, the effective valence should have a value of 0.5, and, under high ionic strength, spermine would appear to the negatively charged channel essentially as an array of discrete positively charged, monovalent nitrogen atoms capable of inducing hydrogen bonds through their protons to polar groups lined at the pore interior.

<u>Calcium currents</u>: The geometry of cell L-6 was also approximated by a prolate spheroid, with a resulting cell volume of 27 nl. The ionophoretic current of 200 nA generated an average concentration of 1.57 mM at 4 min and up to 4.7 mM after 12 min. The upper part of Fig.3 again shows a fit by Eq.2 of experimental values of normalized calcium inward currents and voltages after spermine injection. As the analysis is concerned with the block of open calcium channels, only a narrow range of data along the voltage

axis can be taken into account. In the lower voltage range, the process of activation of calcium channels is superimposed to the ion transport through the open channels, and at higher voltages, nonspecific currents may overlap (Kostyuk, Mironov & Doroshenko, 1982). The parameters as found from fitting Eq.2 to the experimental data are: the binding constant for the first spermine cation, $K_b=(K^0)^{1/n}=260/M$; $n=1.58$; $[SPM_nR]=0.11$ mM; $z\delta=0.8$.

CONCLUSION: Intracellular spermine cations reduce amplitudes and alter the shape of action potentials of bursting pacemaker nerve cells. Also delayed potassium outward currents and calcium inward currents are reduced in a dose dependent manner. The block of the delayed potassium outward currents can be described by a voltage dependent reaction where one spermine cation binds to one channel with the apparent binding constant, $K_b=140/M$. The effective valence, $z\delta=0.45$, implies that, of the 4 positive charges carried by the spermine cation, only one can interact with the electric field within the cell membrane. Thus one charged site may partially penetrate the cell membrane, but the other 3 are excluded. The block of the calcium inward currents can be described by a voltage dependent reaction where more than one spermine cation may interfere with one channel protein. The coefficient, n, that absorbs molecularity and cooperativity of the reaction has a nonintegral value of 1.58. The apparent binding constant of the first spermine cation has the value, $K_b=260/M$. The effective valence, $z\delta=0.8$, for the voltage dependence of the blocking reaction suggests that only one charge of each spermine cation interacts with the electric field within the cell membrane and thus opposes the influx of twovalent calcium ions.

Acknowledgments: The initial experiments of this research were supported by the National Institutes of Health, grant NS 11429-09, and by the Deutsche Forschungsgemeinschaft, grant SFB-156.

REFERENCES

Drouin, H. (1983) In: Membrane Permeability: Experiments and Models. (A.H. Bretag, Ed.), Techsearch Inc., Adelaide, South Australia, pp. 89-96.
Drouin, H. (1984) Biophys. J. 46, 597-604.
Drouin, H. & Hermann, A. (1984) Ann. NY. Acad.Sci. 435, 534-536.
Gilad, G.M. & Gilad, V.H. (1988) Dev. Brain Res. 38, 175-181.
Gorman, A.L.F. & Hermann, A. (1982) J. Physiol. (Lond.) 333, 681-699.
Gould, R.M. & Cottrell, G.A. (1974) Comp. Biochem. Physiol. 48B, 591-597.
Hermann, A. & Gorman, A.L.F. (1981) J. Gen. Physiol. 78, 63-86.
Kandel, E.R. (1976) In: Cellular Basis of Behavior. Freeman.
Kostyuk, P.G., Mironov, S.L. & Doroshenko, P.A. (1982) J. Membrane Biol. 70, 181-189.
Kremzner, L.T. & Ambron, R.T. (1982) J. Neurochem. 38, 1719-1727.
Miller, C. (1982) J. Gen. Physiol. 79, 869-891.
Pajunen, A.E.I., Hietala, O.A., Virransalo, E.L. & Piha, R.S. (1978) J. Neurochem. 30, 281-283.
Schuber, F. (1989) Biochem. J. 260, 1-10.
Slotkin, T.A. & Bartolome, J. (1986) Brain Res. Bull. 17, 307-320.

SPECTROSCOPIC STUDIES ON THE STRUCTURE AND AGGREGATION OF CALCIUM IONOPHORE, A23187

K.R.K. Easwaran and S.V. Balasubramanian

Molecular Biophysics Unit, Indian Institute of Science, Bangalore 560 012, India.

SUMMARY: Studies on the structure of calcium ionophore, A23187, using Infra-red, Ultraviolet, Circular dichroism, Fluorescence and Nuclear magnetic resonance methods indicated that this ionophore forms a dimer in non-polar solvents such as chloroform. When incorporated into a phospholipid vesicle, the spectra showed remarkable time dependent changes indicating further aggregation of the dimer into a stacked dimeric structure in DPPC vesicle. Results point to the possibility of A23187 transporting divalent cations across membranes by a pore rather than by a diffusive carrier model.

The transport of cations across biological membranes is an important process generally mediated by membrane integral proteins (Giebisch et al, 1980; Scarpa & Azzone, 1970). However, understanding of the mechanism of ion transport at the molecular level has essentially been due to the discovery of small macrocyclic and linear antibiotics, ionophores, which are compounds isolated from bacterial sources and which selectively enhance the ionic permeability across model and biological membranes (Pressman, 1976; Bakker, 1979; Ovchinnikov et al., 1974; Pressman, 1967; Andreoli et al., 1967; Mueller & Rudin, 1967; Szabo et al., 1969). During the last few years several physico-chemical and spectroscopic studies have been reported on the structure, ion-complexing ability of the ionophores and their interaction with model and biological membranes (Ovchinnikov et al., 1974; Ivanov, 1975; Grell et al., 1973; Grell et al., 1974; Easwaran, 1985; Easwaran, 1987). In this paper we report our studies on one of the important naturally occurring ionophore, namely, A23187 (also called Calcimycin) which facilitates selective transport of calcium ions across biological membranes (Pfeiffer et al., 1974; Reed & Lardy, 1972).

A23187 which is isolated from cultures of <u>streptomyces</u> <u>chartreu-</u><u>sensis</u>, has a structure consisting of a keto pyrrole group, a spiroketal portion and a benzoxazole ring (Figure 1). The free form and its divalent

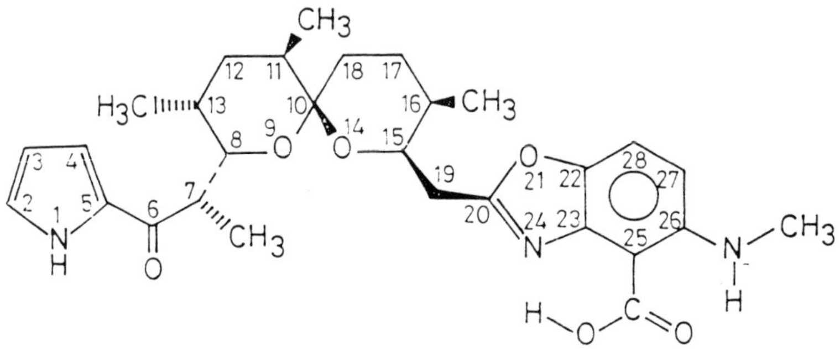

Figure 1: The chemical structure of A23187

complex has been studied in solution by Ultraviolet (UV), Circular dichroism (CD), Fluorescence and Nuclear magnetic resonance (NMR) spectroscopy (Pfeiffer et al., 1974; Divakar & Easwaran (1987); Balasubramanian & Easwaran, 1989; Puskin et al., 1981; Anteunis, 1977). The detailed NMR studies on the solution conformation of A23187 and its Mg^{2+} ion complex in $CDCl_3$ indicated that the free A23187 exists as a monomer in solution (Anteunis, 1977) and is similar to that reported for its structure in the solid state (Chaney et al., 1974). In the Mg^{2+} and Ca^{2+} complex it has been shown that two molecules of A23187 aggregate around the metal cation forming a 2:1 dimer complex with a C_2 symmetry and a cavity of about 3.5 $\overset{o}{A}$ diameter (Smith & Duax, 1976). Fluorescence studies on the free acid and its Mg^{2+} complex showed a shift in the position of emission maxima to shorter wavelength, decreased intensity and alteration of the excitation spectra in the complex as compared to the free acid (Pfeiffer et al., 1974). Fluorescence spectral observation in natural membrane and phospholipid vesicles (Case et al., 1974; Puskin et al., 1981) showed enhanced quantum yield relative to that in aqueous phase and analysis of the data indicated the formation of an equimolar complex near the membrane water interface and 2:1 ionophore:cation complex in the interior of the membrane (Casewell & Pressman, 1972; Kobler & Haynes, 1981).

Our results reported in this paper on the structure and conformation of A23187 as studied by Infra-red (IR), UV, CD, Fluorescence and NMR methods indicated that the free molecule as well as its divalent complex exists as a dimer in non-polar solvents and that they aggregate into a stacked dimeric pore structure in phospholipid vesicle.

MATERIALS AND METHODS

A23187 and dipalmitoyl phosphatidyl choline (DPPC) are from Sigma Chemical Company. The organic solvents used are of spectroscopic grade. The unilamellar vesicles (ULV's) of DPPC were made by sonicating the multilamellar vesicles (MLV's) which are prepared by dispersing in an aqueous medium a thin film of DPPC in a round bottomed flask.

The IR spectra were run on a Bruker FT-IR spectrometer, the UV on a Hitachi double beam UV spectrophotometer, the CD on a Jasco J-500 spectropolarimeter, the fluorescence measurements on a Perkin-Elmer 44A fluorescence spectrophotometer and NMR on a Bruker WH-270 FT NMR spectrometer.

RESULTS AND DISCUSSION

<u>Studies in chloroform</u>: The infra-red spectra of A23187 in chloroform taken at different concentrations showed significant changes in some of the IR bands (only the region 1550-1800 cm^{-1} is shown in Figure 2). The broad band around 1730 cm^{-1} at lower concentration of the ionophore (Figure 1a) corresponds to the carboxyl C=O and the band of small intensity around 1700 cm^{-1} corresponds to the ketopyrrol carbonyl group. As the concentration of the ionophore is increased (Figure 1b) an additional peak around 1650 cm^{-1} was observed. The peak around 1730 cm^{-1} decreased in intensity and the peak at 1700 cm^{-1} increased in intensity and unshifted. Further increase in the concentration (Figure 1c) resulted in the disappearance of the band at 1730 cm^{-1} and increase in the intensity of peaks at 1700 cm^{-1}

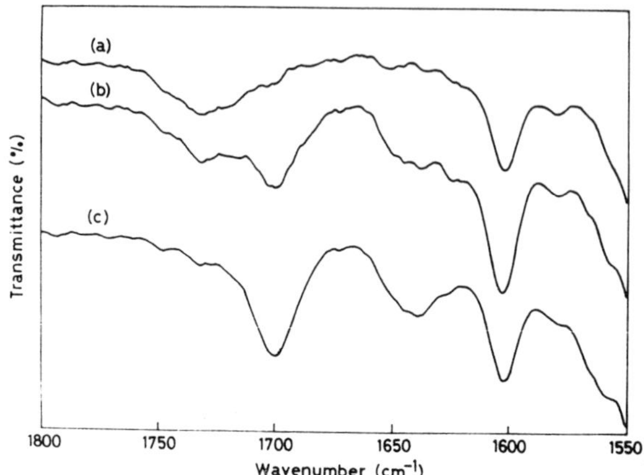

Figure 2:   IR spectra of A23187 in CHCl$_3$.   Concentration of A23187 a) 1 x 10^{-4} b) 5 x 10^{-2} M   c) 1 x 10^{-1} M.

and 1650 cm^{-1}.   The results show that as the concentration of the ionophore is increased the position of the ketopyrrole carbonyl peak is unchanged but the carboxyl C=O stretching band is shifted from about 1730 cm^{-1} to 1650 cm^{-1} indicating that at high concentration the carboxyl C=O is hydrogen bonded to the N–H group of the ketopyrrole group of an adjacent molecule resulting in a dimer.

The UV spectrum of free A23187 in CHCl$_3$ displayed four resolvable bands at 380, 294, 280 and 245 nm.   The peak intensity increased with increasing concentration and saturated beyond a concentration of 1.45 x 10^{-4} M.   Saturation of the peak intensity at higher concentration is possibly due to the formation of a dimer.   The CD spectrum of A23187 in chloroform gave four bands corresponding to the UV absorption peaks.   The spectra showed large changes in intensity with concentration and time which indicated that the molecule has a tendency to aggregate as a dimer in non-polar solvents even in the absence of cations (Balasubramanian & Easwaran, 1989).

The fluorescence spectrum of A23187 in chloroform also showed significant quenching of the fluorescence intensity with time.   This again points to the evidence of a dimer formation.   The final spectrum was very similar to that obtained for the spectra of A23187-divalent cation complex.

The proton NMR spectra of A23187 and its complexes with various cations were recorded at 270 MHz. The assignments of the resonances were done using 2-dimensional COSY spectrum. A typical 270 MHz COSY spectrum of A23187 in $CDCl_3$ is shown in Figure 3. Careful analysis of the spectra and the NOE data (Table) and model building showed that the data could only be interpreted using a dimeric structure for the molecule. The dimer formation is being stabilized by the formation of a hydrogen bond between NH of ketopyrrole group of

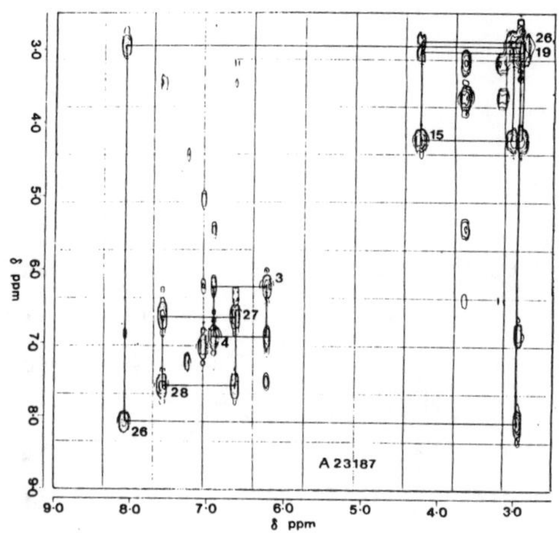

Figure 3: 270 MHz ^{1}H NMR COSY spectrum of A23187 in $CDCl_3$. Concentration of A23187 is 25 mM.

Table: Proton NOE data for A23187 in $CDCl_3$

Proton irradiated[a]	Proton for which NOE observed[a]
26 (NH Me)	25 (COOH), 11, 12
11, 12	26 (NH Me), 25 (COOH)
28	27, 11, 12
15	8

[a] see Figure 1 for the numbering scheme

one molecule and the carbonyl group of the second molecule. The chemical shift and coupling constant changes observed in the NMR spectra of A23187-Mg^{2+} complex could be explained as due to the changes induced by the divalent cation co-ordinating to the carboxyl oxygen, a carbonyl oxygen and nitrogen of the benzoxazole ring system of each ionophore molecule in the dimeric structure. Our results on the solution conformation of A23187 in CDCl$_3$ showing a dimeric structure is in disagreement with that reported earlier wherein it was suggested that the molecule exists as a monomer (Anteunis, 1977).

Studies of A23187 in DPPC vesicle:

The CD spectra recorded for A23187 incorporated into DPPC ULV's and taken within minutes after sonication was similar to that obtained in chloroform at high concentration of the ionophore or after several hours at lower concentration, except for an additional peak of small intensity around 345 nm. However, the spectra showed remarkable changes with time and stabilized after about 40 minutes. Data shows clearly that the molecule aggregates readily into a dimer in a lipid environment and with time the dimer molecule further aggregates into a stacked dimeric structure with NH's of the benzoxazole ring in one dimer hydrogen bonded to the C=O of the ketopyrrole group of the adjacent dimeric molecule.

The fluorescence spectrum of A23187 in DPPC vesicle also exhibited decrease in fluorescence intensity with time (Figure 4). The fluorescence polarization value for A23187 in DPPC (0.43) were an order of magnitude larger than that observed in CHCl$_3$ (0.05). These results also support the fact that A23187 which has a dimeric structure in chloroform aggregates further in a lipid vesicle. The proposed model of stacked dimeric structure for A23187 in phospholipid vesicle is given in Figure 5.

CONCLUSION: Our results on the studies on the structure of A23187 using variety of physico-chemical and spectroscopic techniques clearly showed that the molecule exists as a dimer in non-polar solvents such as chloroform and aggregates into a stacked dimeric pore in phospholipid vesicle. This leads to the possibility of calcium transport

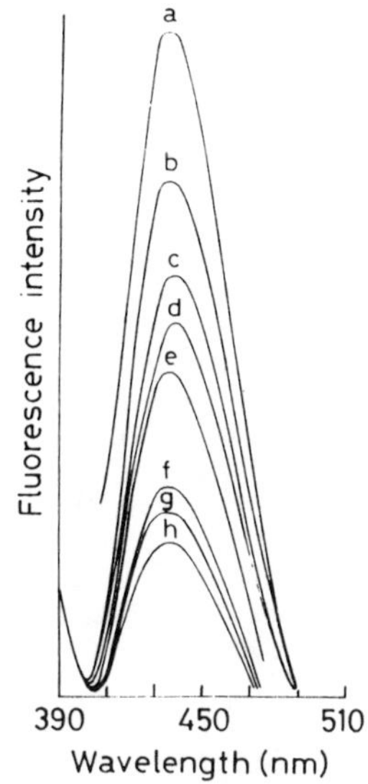

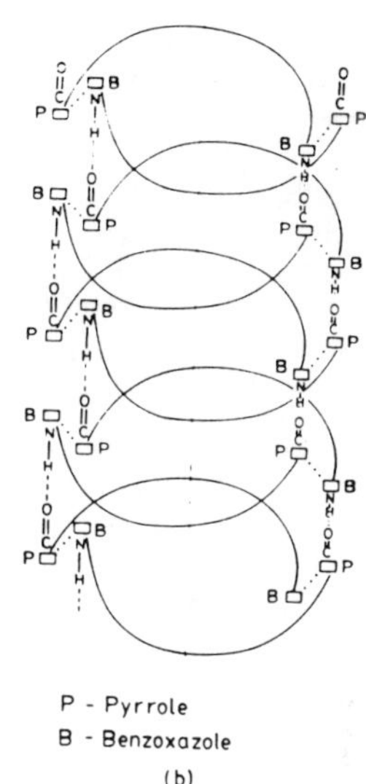

Figure 4: Fluorescence spectra of A23187 in DPPC vesicle as a function of time. Concentration of A23187 is 1.6×10^{-12}M; lipid:ionophore is 2000:1; a) 0 time b) 45 mts. c) 1.45 hrs d) 2.45 hrs e) 3.45 hrs f) 20 hrs g) 22 hrs h) 26 hrs.

Figure 5: Proposed model of stacked dimeric pore structure for A23187 in phospholipid vesicle.

by this ionophore across membrane by a pore rather than by a diffusive carrier mechanism. The proposed model will also explain the large turnover number for calcium by A23187, as compared to other ionophores of similar structure, in membranes as reported in the literature (Casewell & Pressman, 1972).

REFERENCES

Andreoli, T.E., Tieffenberg, M. and Tosteson, D.C. (1967) J.Gen.Physiol. 50, 2527

Anteunis, M.J.O. (1977) Bioorg.Chem. 6, 1-11

Bakker, E.P. (1979) "Antibiotics" (F.E. Hahn, Ed.) Vol. V, Springer-Verlg, Berlin.

Balasubramanian, S.V. and Easwaran, K.R.K. (1989) Biochem.Biophys. Res.Comm. 158, 891-897

Case, G.D., Vander Kooi, J.M. and Scarpa, A. (1974) Arch.Biochem. Biophys. 162, 174-185

Casewell, A.H. and Pressman, B.C. (1972) Biochem.Biophys.Res.Comm. 49, 292-298

Chaney, M.O., Demaro, P.V., Jones, N.D. and Occolowitz, J.C. (1974) J.Am.Chem.Soc. 96, 1932-1933

Divakar, S. and Easwaran, K.R.K. (1987) Biophys.Chem. 27, 139-147

Easwaran, K.R.K. (1985) "Metal ions in Biological Systems" (H. Sigel, Ed.) Vol. 19, Chap.5, Marcel Dekker, New York

Easwaran, K.R.K. (1987) "Ion transport through Membranes" (Kunio Yagi and Bernard Pullman, Eds), Chap.2, Academic Press, Japan

Giebisch, G., Tosteson, D.C. and Ussing, H.H. (1980) "Membrane transport in Biology", Vol. III, Springer-Verlag, Berlin.

Grell, E., Funck, T. and Sauter, H.P. (1973) Eur.J.Biochem. 34, 415-

Grell, E., Funck, T. and Eggers, F. (1974) "Membranes, A Series Advances" (G. Eisenman, Ed), Vol. III, Chap.1, Marcel Dekker, New York.

Ivanov, V.T. (1975) Ann.N.Y.Acad.Sci. 264, 221-243.

Kolber, M.A. and Haynes, D.H. (1981) Biophys.Jour. 36, 369-390.

Mueller, P. and Rodin, D.O. (1967) Biochem.Biophys.Res.Comm. 26, 398-404

Ovchinnikov, Yu.A., Ivanov, V.T. and Shkrob, A.M. (1974) "Membrane active complexones", Elsevier/North Holland, Amsterdam.

Pfeiffer, D.R., Reed, R.W. and Lardy, H.A. (1974) Biochem. 13, 4007-4013.

Pfeiffer, D.R., Taylor, R.W. and Lardy, H.A. (1978) Ann.N.Y.Acad.Sci. 307, 402-421.

Pressman, B.C., Harris, E.J., Jagger, W.S. and Johnson, J.H. (1967) Proc.Natl. Acad.Sci. U.S.A. 58, 1949-1956.

Pressman, B.C. (1976) Ann.Rev.Biochem. 45, 501-530.

Puskin, J.S., Visties, M.I. and Wene, M.T. (1981) Arch.Biochem.Biophys. 206, 164-172.

Reed, R.W. and Lardy, H.A. (1972) J.Biol.Chem. 247, 6970-6977.

Scarpa, A. and Azzone, G.F. (1970) Eur.J.Biochem. 12, 328

Smith, G.D. and Duax, W.L. (1976) J.Am.Chem.Soc. 98, 1578-1580.

Szabo, G., Eisenman, G. and Ciani, S. (1969) J.Membrane Biol. 1, 346-382.

FURA-2 IMAGING OF INTRACELLULAR FREE CALCIUM DYNAMICS IN EXCITABLE CELLS

M. Grouselle and D. Georgescauld

Centre de Recherche Paul Pascal, Avenue A. Schweitzer, 33600 Pessac, France

SUMMARY: Intracellular free Ca^{2+} concentration ($[Ca^{2+}]_i$) maps were obtained from excitable cells loaded with fura-2, using improved digital image fluorescence microscopy. New procedures allowed to get fast sampling of $[Ca^{2+}]_i$ maps (25 maps/sec) and real spatial distribution of $[Ca^{2+}]_i$ changes only. The spatial heterogeneity of $[Ca^{2+}]_i$ changes in skeletal muscle cells stimulated by KCl depolarisation is interpreted as a non-uniform distribution of functional sarcoplasmic reticulum.

Recent technological advances concerning the synthesis of membrane permeant fluorescent Ca^{2+} indicators (Grynkiewicz *et al.*, 1985) and the digital fluorescence microscopy (Tsien and Poenie, 1986) allow to continuously monitor the intracellular free calcium concentration ($[Ca^{2+}]_i$) and to follow the spatial distribution of $[Ca^{2+}]_i$ over the time course of transients.

Detailed studies of spatio-temporal distribution of $[Ca^{2+}]_i$ have shown that stimulated excitable cells as neurons (Connor, 1986; Connor *et al.*, 1987; Tank *et al.*, 1988 Lipscombe *et al.*, 1988; Lipscombe *et al.*, 1988; Connor *et al.*, 1988; Connor and Tseng, 1988), cardiac muscle cells (Wier *et al.*, 1987; Cheung *et al.*, 1989; Tamura *et al.*, 1989; Geerts *et al.*, 1989), smooth muscle cells (Williams , *et al.*, 1985; Williams *et al.*, 1987; Erne and Hermsmeyer, 1988; Goldman and Blaustein, 1988) or glandular cells (Connor *et al.*, 1987; O'Sullivan *et al.*, 1989) presented a remarkable spatial organisation of the $[Ca^{2+}]_i$ distribution.

If $[Ca^{2+}]_i$ transients in skeletal muscle fibres are well documented (for references see Brum *et al.*, 1988), so far, no investigations were undertaken to explore spatial and temporal Ca^{2+} profiles within single isolated skeletal muscle cells. It was already suggested that the $[Ca^{2+}]_i$ increase in a stimulated skeletal muscle cell may be non-uniformly distributed, the highest concentration being achieved immediatly around the sites of Ca^{2+} release, presumably the terminal cisternae of the sarcoplasmic reticulum (Blink *et al.*, 1978).

In this article, it is shown for the first time, the mapping of $[Ca^{2+}]_i$ transients in fura-2 loaded myotubes stimulated by KCl depolarisation, using digital image fluorescence microscopy improved by new procedures for fast sampling (every 40 msec) of $[Ca^{2+}]_i$ maps and for detection of real spatial distribution of $[Ca^{2+}]_i$ changes.

MATERIALS AND METHODS

Fura-2 imaging: briefly, an Olympus IMT-2 inverted microscope equiped with two excitation filters (350 nm and 380 nm) and an emission long pass filter (> 470 nm), a SIT camera, a photomultiplier, a video cassette recorder U-matic and a video monitor composed the imaging system. The recorded frames are digitized and analyzed with an image processor Pericolor 2001 connected to a VAX 8600 computer. The details of the optical system, the loading of cells with the Ca^{2+} indicator, the image processing and the calibration of $[Ca^{2+}]_i$ are described elsewhere (Grouselle *et al.*,

1989). The free Ca^{2+} concentration is measured from the image ratio (R) obtained by dividing on pixel-by-pixel basis the fluorescence image obtained at 350 nm excitation wavelength by the fluorescence image collected at 380 nm excitation and according to the equation (Grynkiewicz *et al.*, 1985):

$$[Ca^{2+}]_i = \beta \; Kd \; \frac{R - R_{min}}{R_{max} - R} \qquad (1)$$

where R represents the 350/380 nm fluorescence image ratio of the fura-2 loaded cell, R_{min} is the minimum value of the 350/380 nm fluorescence ratio in the absence of calcium (0 mM Ca $^{2+}$), R_{max} is the maximum value of the same ratio at saturating levels of Ca^{2+} (> 1 mM Ca^{2+}), β is the ratio of the 380 nm excitation fluorescence in the absence and saturating levels of Ca^{2+} respectively and Kd is the dissociation constant of the fura-2-Ca^{2+} reaction (224 nM). The calibration parameters obtained on the myotubes were as follows: R_{max} = 10.8, R_{min} = 0.46, β = 10.2.

<u>Fast sampling of ratio images</u>: a pair of fluorescence images in order to calculate the ratio image is obtained every 2-3 seconds due to mechanical limitations of the filters change. For fast sampling of ratio images, the fluorescence image collected at 350 nm excitation from a cell in resting conditions is divided by each frame recorded at video rate (25 frames/sec) at 380 nm excitation during the time course of a $[Ca^{2+}]_i$ transient. This procedure is supported by the following arguments: a) the 350 nm excitation wavelength being very close to the isosbestic wavelength of fura-2, the variation of fluorescence intensities are very small, even at high $[Ca^{2+}]_i$ increases; b) the recording of fluorescence images obtained at 380 nm excitation lasted only few seconds, so cellular shape variations and leakage or photobleaching of fura-2 are negligeable.

<u>Spatial analysis of ratio images</u>: for identifying correctly the spatial distribution of $[Ca^{2+}]_i$ changes, the following operation on image ratios were introduced: a) the difference of ratio images $(R_t - R_o)$; b) the normalization of ratio images (R_t/R_o); c) the normalization of the difference of ratio images $((R_t - R_o)/R_o)$ where R_o is the ratio image of the myotube in resting conditions (image reference) and R_t is the ratio image of the cell at time t during the time course of the $[Ca^{2+}]_i$ transient. Considering the equation (1)

$$R = \frac{[Ca^{2+}]_i\ R_{max}\ +\ \beta\ Kd\ R_{min}}{\beta\ Kd\ +\ [Ca^{2+}]_i} \tag{2}$$

If $[Ca^{2+}]_i^{t=t}$ and $[Ca^{2+}]_i^{t=0}$ represent $[Ca^{2+}]_i$ at time t and at time 0 respectively and if it is considered that $\beta\ Kd = 2285$ nM is much higher than $[Ca^{2+}]_i^{t=t}]$ and $[Ca^{2+}]_i^{t=0}$, it is easy to demonstrate that $R_t - R_o$ is proportional to $[Ca^{2+}]_i^{t=t} - [Ca^{2+}]_i^{t=0}$ whereas R_t/R_o and $(R_t - R_o)/R_o$ are linearly related to $[Ca^{2+}]_i^{t=t} / [Ca^{2+}]_i^{t=0}$ and $[Ca^{2+}]_i^{t=t} - [Ca^{2+}]_i^{t=0}$ respectively. These procedures allow to obtain maps of $[Ca^{2+}]_i$ changes, minimizing the possible artifacts due to the optical system and to cellular preparation.

<u>Cell cultures</u>: Myotubes were prepared from the hind legs of 18 day-old rat embryos as previously described (Koenig *et al.*, 1982). For fura-2 loading and fluorescence measurements, myotubes were maintained in Hepes buffered solution (HBS), pH 7, containing (in mM): NaCl 118, KCl 4.6; $CaCl_2$ 1.8; $MgCl_2$ 1; D-glucose 10; Hepes 20.

RESULTS

In myotubes loaded with fura-2, resting $[Ca^{2+}]_i$ averaged 82 $\pm$ 0.06 nM, (n=46) but, for individual cells, resting $[Ca^{2+}]_i$

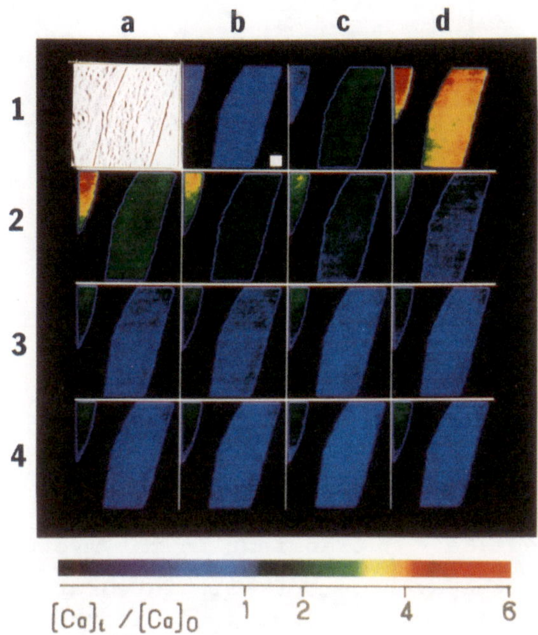

Fig. 1 Sequence of normalized ratio images (R_t/R_0) illustrating a $[Ca^{2+}]_i$ transient induced by 60 mM KCl in a five-day-old rat myotube. The pictures follow one another from top row and from left to right. Rectangular coordinates expressed by numbers and letters indicate a given image in the sequence. (1a): phase-contrast image of the myotubes; image dimension 25×25 μm; (1b): normalized ratio image in resting conditions (R_0/R_0) which show the reference level at t=0. For every following picture is given the time t after the starting of KCl depolarization. (1c): t=5 sec; (1d): t=13 sec; (2a): t=21 sec; (2b): t=29 sec; (2c): t=38 sec; (2d): t=48 sec; (3a): t=55 sec; (3b): t=65 sec; (3c): t=80 sec; (3d): t=90 sec; (4a): t=108 sec; (4b): t=123 sec; (4c): t=150 sec; (4d): t=180 sec. Note the highest $[Ca^{2+}]_i$ increase in the picture (1d) in both myotubes shown in the center and in the left top corner of the image. Image dimension 30×30 μm. The color bar is calibrated in $[Ca^{2+}]_i^{t=t} / [Ca^{2+}]_i^{t=0}$.

234

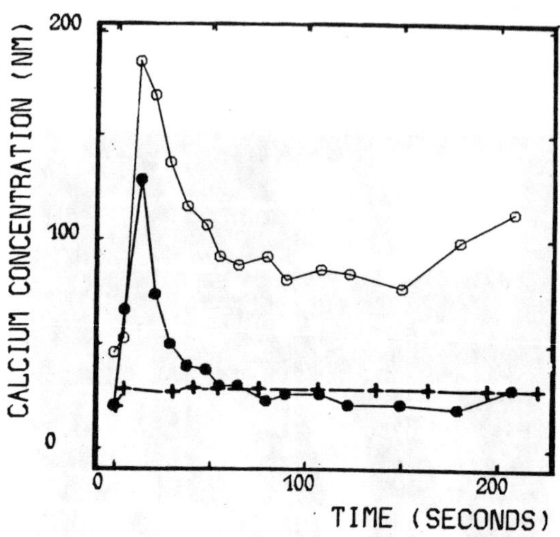

Fig. 2 KCl induced $[Ca^{2+}]_i$ transient in rat myotubes. Data points represent the average of all integrated $[Ca^{2+}]_i$ values. The first 14 points correspond to the pictures (1b) to (4d) shown in Fig. 1 (•) central myotube in Fig. 1, (o) left top corner myotube in Fig. 1, (+) block of $[Ca^{2+}]_i$ transient in the presence of D-600 100 µM.

are quite variable. Fig. 1 presents an example of the $[Ca^{2+}]_i$ transient induced by 60 mM KCl depolarization in two myotubes. The sequence of pictures demonstrate the improvment provided by the normalized ratio image procedure which allows to measure accurately only changes of $[Ca^{2+}]_i$. In the images (1d), (2a) and (2b) the local increases in $[Ca^{2+}]_i$ during the transient in both myotubes may vary by a factor as high as 3, as areas of red in both cells indicate locally large increments of Ca^{2+} activity. The rise of $[Ca^{2+}]_i$, about four times the basal level, takes less than 15 sec and the recovery of the resting $[Ca^{2+}]_i$ level is obtained in about 35 sec, in spite of the continuous presence of high KCl concentration. The effect of KCl on $[Ca^{2+}]_i$ transients is reproducible and the time courses of transients are similar

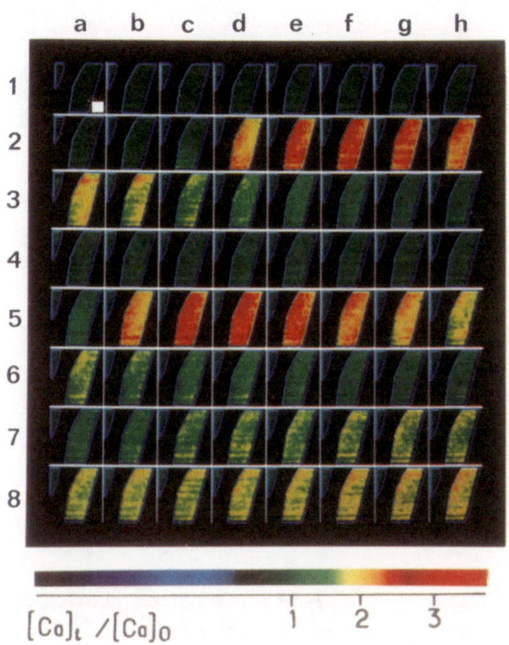

Fig. 3 Sequence of normalized calcium variations images (R_t / R_0) obtained every 40 msec from the central rat myotube shown in Fig. 1 and illustrating two very early fast $[Ca^{2+}]_i$ transients induced by 60 mM KCl. The time course of the $[Ca^{2+}]_i$ transients starts from the top row and from left to right and a given picture is indicated as described in the legend of Fig. 1. Note the $[Ca^{2+}]_i$ fast oscillations shown in the pictures from (2d) to (3d) and from (5b) to (6b) and the non-uniform spatial distribution of $[Ca^{2+}]_i$ changes. The $[Ca^{2+}]_i$ of the myotube shown on the left top corner of the pictures remained at its basal level during the first two seconds of the $[Ca^{2+}]$ transients. Note also the very different time scale between Fig. 1 and Fig. 3. Image dimension 30 × 30 μm. The color bar is calibrated in $[Ca^{2+}]_i^{t=t} / [Ca^{2+}]_i^{t=0}$.

even when the basal $[Ca^{2+}]$ of myotubes are quite different (Fig. 2). The fast sampling procedure joined to the normalization of ratio images may detect in some myotubes faster and smaller

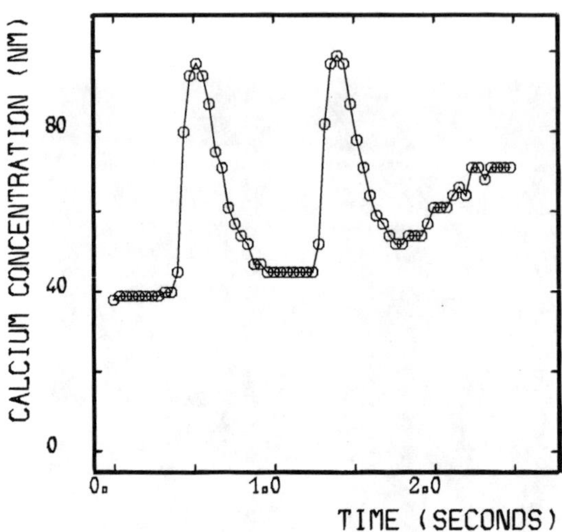

Fig. 4 Very early fast $[Ca^{2+}]_i$ transients induced by 60 mM KCl in the central myotube shown in Fig. 3. Data points represent the average of integrated $[Ca^{2+}]_i$ values obtained from the myotube area. The two $[Ca^{2+}]_i$ oscillations correspond to the pictures (2d)-(3d) and (5b)-(6b) shown in Fig. 3.

$[Ca^{2+}]_i$ oscillations which precede the main and slower $[Ca^{2+}]_i$ transient (Fig. 3). So, it was possible to demonstrate that the pattern of the very beginning of KCl induced $[Ca^{2+}]_i$ transient show a different behaviour during the first two seconds after the start of KCl depolarisation. It is striking that the myotube situated in the center of the field (Fig. 3) presents fast $[Ca^{2+}]_i$ oscillations whereas the $[Ca^{2+}]_i$ of the myotube observed at the top left corner of the pictures in Fig. 3 stays at its basal level during the same time interval. The rising time of the fast $[Ca^{2+}]_i$ transients is about 160 msec and the maximum average increase may reach only 2.5 times the basal level of $[Ca^{2+}]_i$ (Fig. 4). As it is seen in the pictures from (2d) to (3d) and from (5b) to (6b) presented in Fig. 3, the spatial distribution of $[Ca^{2+}]_i$

changes is clearly non-uniform. It may be noticed that KCl induced $[Ca^{2+}]_i$ transients are completely abolished by 100 μM of D-600, a specific blocker of slow Ca^{2+} channels (Fig. 2).

DISCUSSION

Myotubes possess spontaneous and induced propagated action potentials, functional acetylcholine receptors and contractility, all properties close to those of denervated adult muscle fibres (Schmid-Antomarchi *et al.*, 1985). Recently, detailed electrophysiological and pharmacological properties of two classes of potential-dependent Ca^{2+} channels expressed in rat myotubes have been reported (Cognard *et al.*, 1986; Beam and Knudson, 1988), and it was well established that the excitation-contraction coupling mechanisms in myotubes and adult skeletal muscle cells are very similar (Romey *et al.*, 1988).

Our results show that the $[Ca^{2+}]_i$ responses of myotubes depolarized by high K^+ are diverse (Fig. 3) and we suggest that this variability may represent different developmental states of the myotubes in culture as it was already discussed from electrophysiological data (Land *et al.*, 1973; Beam and Knudson, 1988).

The main $[Ca^{2+}]_i$ transients in the presence of high K^+ may represent the release and the reuptake of Ca^{2+} by the sarcoplasmic reticulum as voltage-champ analysis associated to contraction measurements have clearly shown that Ca^{2+} influx through L-type Ca^{2+} channels is not required for contraction (Romey *et al.*, 1988) and as the time courses of the Ca^{2+} action potential and of the $[Ca^{2+}]_i$ transients are very different (50 msec v.s. 50 sec respectively).

A non-homogeneous distribution of diads and triads which are believed to be responsible for the excitation-contraction coupling between the membrane depolarization and the Ca^{2+} release from the sarcoplasmic reticulum (Caswell and Brandt, 1989) may explain the striking finding concerning the reproducible non-uniform distribution of $[Ca^{2+}]_i$ transients (Fig. 1 and Fig. 3). This hypothesis is supported by electron microscopic data which have shown that, in rat myotubes, couplings between the sarcoplasmic reticulum and the plasmalemma or the few T tubules are rare and inhomogeneously distributed (Kelly, 1971).

The fast $[Ca^{2+}]_i$ oscillations, recorded sometimes at the very begining of the high K^+ depolarization (Fig. 3 and Fig. 4) may have a different origin, presumably linked to the $[Ca^{2+}]_i$-dependent-K^+ channel which is supposed to control the pacemaker activity of myotubes and denervated adult muscle fibres (Cognard *et al.*, 1986).

The abolishment of the $[Ca^{2+}]_i$ transient by a quite high concentration of D-600 (100 μm), a specific L-type Ca^{2+} channel blocker, does not mean a need for Ca^{2+} ions entry to start the $[Ca^{2+}]_i$ changes but instead the cancellation of the link between membrane depolarization and the mean components of charge movements (Hui *et al.*,, 1984). Despite the high D-600 concentration, we have never seen a $[Ca^{2+}]_i$ increase suggested by the hypothetical local anaesthetic effect of D-600 on the sarcoplasmic reticulum (Dörrscheidt-Käfer, M., 1977).

CONCLUSION: The technique of fast sampling of ratio images coupled to the normalization procedure improved markedly the time resolution and the confidence in the real spatial distribution of $[Ca^{2+}]_i$ changes in stimulated excitable cells.

Acknowledgments: We thank J. Koenig and J. Chapron for providing the rat myotubes. Financial supports from Ministère de la Recherche et de la Technologie, Conseil Régional de l'Aquitaine and As-

REFERENCES

socition Française pour la lutte contre les Myopathies are kindly
acknowledged.

Beam, K.G., and Knudson, C.M. (1988).
 J. Gen. Physiol., 91, 781-798.
Blinks, J.R., Rüdel, R., and Taylor, S.R. (1978)
 J. Physiol. (Lond.), 277, 291-323.
Brum, G., Rios, E., and Stefani, E. (1988)
 J. Physiol. (Lond.), 398, 441-473.
Caswell, A.H., and Brandt, N.R. (1989)
 TIBS, 14, 161-165.
Cheung, J.Y., Tillotson, D.L., Yelamarty, R.V., and Scaduto Jr,
 R.C. (1989)
 Am. J. Physiol., 256(25), C1120-C1130.
Cognard, C., Lazdunski, M., and Romey, G. (1986)
 Proc. Natl. Acad. Sci. USA, 83, 6179-6183.
Connor, J.A., Tseng, H.Y., and Hockberger, P.E. (1987).
 J. Neurosci. 7(5), 1385-1400.
Connor, J.A., Carter Cornwell, M., and Williams, G.H. (1987).
 J. Biol. Chem., 262, 1919-1927.
Connor, J.A., Wadman, W.J., Hockberger, Ph. E., Wong, R.K.S.
 (1988)
 Science, 240, 649-653.
Connor, J.A., and Tseng, H.Y. (1988)
 Brain Res. Bull., 21, 353-361.
Dörrscheidt-Käfer, M. (1977)
 Pflügers Arch., 369, 259-267.
Erne, P., and Hermsmeyer, K. (1978)
 J. Cardiovasc. Pharmacol., 12(5), S85-S91.
Geerts, H., Nuydens, R., Nuyens, R., and Ver Douch, L. (1989)
 Cardiovas. Res., 23, 797-806.
Goldman, W.F., and Blaustein, M.F. (1988)
 J. Cardiovasc. Pharmacol., 12(5) S13-S19.
Grouselle, M., Koenig, J., Chapron, J., and Georgescauld, D.
 (1989)
 (Submitted)
Grynkiewicz, G., Poenie, M. and Tsien, R.Y. (1985)
 J. Biol. Chem., 260, 3440-3450.
Hui, C.S., Milton, R.L., and Eisenberg, R.S. (1984)
 Proc. Natl. Acad. Sci. USA, 81, 2582-2585.
Kelly, A.M. (1971)
 J. Cell. Biol., 49, 335-344.
Koenig, J., Bournaud, R., Powell, J.A., and Rieger, F. (1982)
 Dev. Biol., 92, 188-196.
Land, B.R., Sastre, A., and Podleski, T.R. (1973)
 J. Cell. Physiol., 82, 497-510.

Lipscombe, D., Madison, D.V., Poenie, M., Reuter, H., Tsien,
     R.W., and Tsien, R.Y. (1988)
     Neuron, 1, 355-365.
Lipscombe, D., Madison, D.V., Peonie, M., Reuter, H., Tsien,
     R.Y., and Tsien, R.W. (1988)
     Proc. Natl. Acad. Sci. USA, 85, 2398-2402.
O'Sullivan, A.J., Cheek, T.R., Moreton, R.B., Berridge, M.J., and
     Burgoyne, R.D. (1989)
     EMBO J., 8, 401-411.
Romey, G., Garcia, L., Rieger, F., and Lazdunski, M. (1988)
     Biochem. Biophys. Res. Commun., 156, 1324-1332.
Schmid-Antomarchi, H., Renaud, J.F., Romey, G., Hugues, M.,
     Schmid, A., and Ladzunski, M. (1985)
     Proc. Natl. Acad. Sci. USA, 82, 2188-2191.
Tamura, K., Yoshida, S., Fujiwake, H., Watanabe, I., and Sagawa-
     ra, Y. (1989)
     Biochem. Biophys. Res. Commun., 162, 926-932.
Tank, D.W., Sugimori, M., Connor, J.A., and Llinas, R.R. (1988).
     Science, 242, 773-777.
Tsien, R.Y., and Poenie, M. (1986)
     TIBS, 11, 450-455.
Wier, W.G., Cannell, M.B., Berlin, J.R., Marban, E., and Lederer,
     W.J. (1987)
     Science, 235, 325-328.
Williams, D.A., Becker, P.L., and Fay, F.S. (1987)
     Science, 235, 1644-1648.

CHEMICALLY DRIVEN PHASE SEPARATION IN BLACK LIPID MEMBRANES

S. Mittler–Neher, J. Spinke, and W. Knoll

Max–Planck–Institut für Polymerforschung, Ackermannweg 10, D–6500 Mainz, FRG

SUMMARY: We studied the coupling of a membrane function (the transport of Cs^+–ions by the pore–forming polypeptide gramicidin) to chemically driven phase changes in black membranes of binary lipid mixtures.

First we established a miscibility gap in the Ca^{++}–concentration–composition phase diagram of mixed lecithin/phosphatidylglycerol membranes: at very low Ca^{++}-concentrations ($c < 10^{-6}$ M) a conductance histogram with a single population is found which indicates a homogeneously mixed membrane. At higher Ca^{++}–concentrations the histograms show a bimodal distribution of pores with different mean conductivities, Λ_i, showing the demixed state of the membrane with two coexisting phases.

Next, we demonstrate that, under certain conditions, one can switch between a homogeneously mixed membrane and a phase separated state also by changing the pH. We interpret this as being caused by the change in the degree of dissociation of the phosphatidylglycerol headgroup.

Finally, two examples for polyelectrolyte (polylysine and polyvinylmethyl-pyridinium) are given.

INTRODUCTION

The understanding of order–function relations in membranes is still a major challenge in biophysical research, for biomembranes as well as at the model membrane level (Sackmann, 1984a). One particular aspect concerns the coupling of the function of integral proteins to the transversal and lateral organization of the lipids in a bimolecular matrix composed, in general, of many components (Sackmann, 1984b).

We address this question by studying a membrane function — the transport of ions mediated by a model protein, here the pore forming polypeptide gramicidin (Hladky & Haydon, 1970; Bamberg & Läuger, 1973) — as a function of the miscibility (phase) behavior of a black membrane (BLM) composed of a binary lipid mixture. The analysis of the single—channel current fluctuations is particularly well—suited for this purpose because a homogeneous distribution of the lipid components shows as a single narrow conductance histrogram whereas a demixed membrane with two coexisting phases can be identified by a bimodal conductance distribution (Knoll et al., 1986).

We focus in the following on fluid—fluid demixing phenomena driven by charged chemical species because they allow for an isothermal induction of (local and transient) changes in the lateral organization of the lipid bilayer.

MATERIALS AND METHODS

Black membranes were prepared according to the method described by Mueller et al. (1982) from 1% (wt/vol) lipid solutions in n—decane(Fluka, purum). Some test experiments were performed with "virtually solvent—free" membranes following the procedure given by Montal & Mueller (1972). 1,2—dioleoyl—sn—glycero—3—phosphatidylcholine (PC) and 1,2—dioleoyl—sn—glycero—3—phosphatidylglycerol (PG) were obtained from Avanti (Birmingham, Alabama) and used without further treatment. The electrolyte solutions contained in all cases 0.5 M CsCl (Merck, p.A. quality). If even traces of divalent ions had to be removed 10^{-4} M ethylenediaminetetraacetate (EDTA) was added. For some experiments $CaCl_2$ (Merck, p.A.) polylysine (PL, Sigma) or polyvinylmethylpyridinium iodide (PVMP, generous gift of J. Zöller, University of Mainz) was added in various concentrations.

Different pH values were adjusted by adding NaOH and HCl and were checked before and after completion of each experiment. The temperature was 22°C. Gramicidin (a commercial mixture of A, B, and C) was added from methanolic stock solutions as needed.

Single channel current fluctuations were recorded as described and analysed with respect to conductance, Λ, and lifetime, τ, in a way that allowed also for the analysis of sub—populations thereby looking for correlations between Λ and τ (Knoll et al., 1986). It was thus possible to derive in a demixed membrane with two different channels not only their mean conductance, $\overline{\Lambda}_i$, but also their mean lifetime, $\overline{\tau}_i$.

RESULTS AND DISCUSSION

<u>Ca^{++}–induced phase separation</u>: In a series of experiments with different ionophores we had shown that PC and PG mix homogeneously over the full range of composition (PG$_{(1-x)}$ PC$_x$, $0 < x < 1$) provided any divalent ions are removed (Mittler–Neher and Knoll, 1989b). If, however, Ca^{++} ions even in concentrations as low as $10^{-5} - 10^{-6}$ M are present (as impurities in 0.5 M CsCl solutions) demixing of equimolar PC/PG mixtures occurs which shows up clearly in the current fluctuations of incorporated gramicidin. An example is given in Figure 1 for a 10^{-4} M CaCl$_2$ containing solution at pH 4. Two types of channels with distinctly different conductance increments can be seen to open and close independently of each other.

Consequently, a bimodal conductance histogram is found in the statistical analysis of these current fluctuations. One example, this time for a 10^{-3} M CaCl$_2$ containing electrolyte solution (pH 6) is presented in Figure 2a). The appearance of two populations unambiguously indicates the presence of two coexisting phases with different composition, x, and hence different influence on the single channel characteristics of gramicidin. This possibility for the simultaneous but distinguishable observation of two types of channels is <u>the</u> key advantage in using pore–forming systems for such studies. Other ionophores like carriers give by far less direct evidence for phase separation phenomena in BLM (Schmidt et al., 1982). If these measurements are performed for a

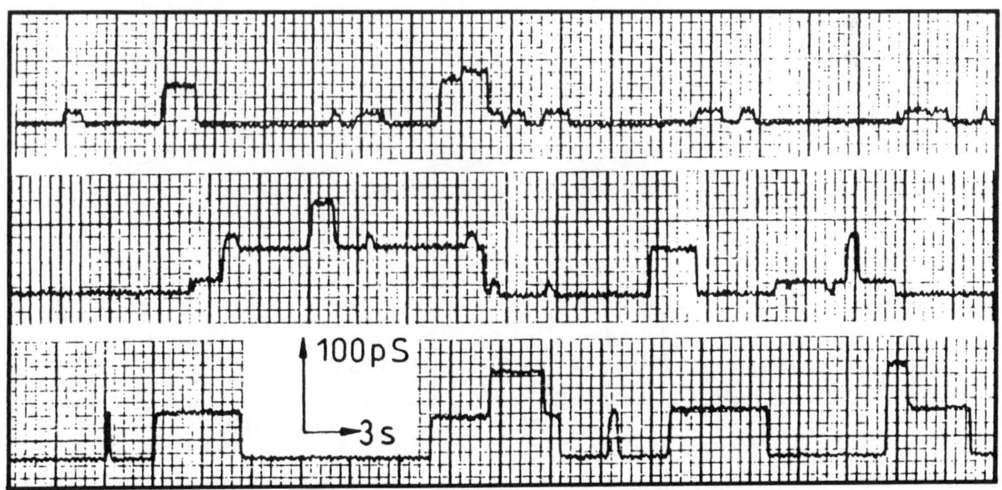

<u>Fig. 1</u>: Record of conductance fluctuations of gramicidin in a demixed membrane (0.5 M CsCl; 10^{-4} M CaCl$_2$; pH 4).

number of mixed membranes with different compositions a fair amount of phase information can be derived for this lipid mixture organized in the form of a single bilayer (Mittler–Neher, 1989). (This then in turn can be compared, e.g., with the miscibility behavior found in vesicles by other techniques (Henkel et al., 1989)). This is demonstrated in Figure 2b). In pure PC–membranes (x = 1) a single channel population is found at $\bar\Lambda = 35$ pS. If some PG is added (x = 0.9) the membrane becomes slightly negatively charged and the channel conductance increases through the enhancement of the interfacial Cs^+–concentration. Further addition of PG now increases the Ca^{++}–concentration (in the bulk 10^{-3} M) at the interface to a level where blocking of the gramicidin by some "binding" to the membrane occurs (Bamberg & Läuger, 1977): the mean conductance increment of the still single channel population decreases. It takes only little more PG ($x \leq 0.85$) to further enhance the interfacial Ca^{++}–concentration and, through its interaction with the charged headgroups, to destabilize the lipid mixtures so that the system undergoes a phase separation – a second channel population appears. It is clear by the lever–rule for coexisting phases that the relative areal fraction of this newly formed phase is rather small. There is, however, a strong tendency for gramicidin to partition preferentially into this PG–rich phase so that the corresponding conductance histogram shows that contribution highly exagerated. The phase boundary of the resulting miscibility gap can thus be determined to certainly better than 5 mole %. This same effect, on the other hand, limits the accuracy for the location of the phase boundary on the PG–rich side of the diagram and may be there not much better

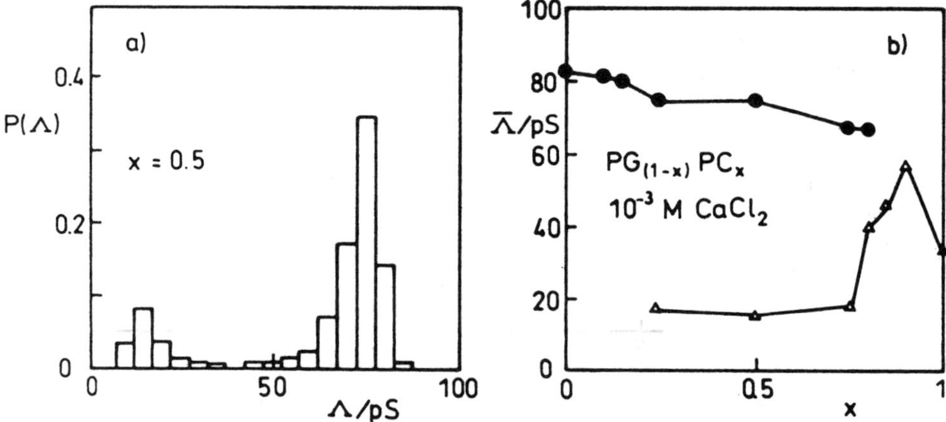

Fig. 2: a) Normalized conductance histrogram P (Λ) of a x = 0.5 membrane; 0.5 M CsCl, 10^{-3} M CaCl₂, pH 6; (b) mean conductance increments, Λ, of a single gramicidin pore in mixed or demixed membranes of various compositions, x. Full circles indicate the PG–rich phase, open triangles the PC–rich phase.

than 10 mole %. As it is expected for a phase separated system the average conductivities of the two coexisting channels are independent of the mole fraction, x, over the whole range of the rather wide miscibility gap.

Further support for this interpretation comes from the analysis of the second characteristic channel parameter, namely its lifetime, τ. One example is given in Figure 3a). Here, it was essential that the two channel populations found in the conductance histogram (Fig. 2a)) were analyzed independently: the pores with conductance increments between 0 and 40 pS (open triangles) live on average considerably shorter than the ones between 40 and 100 pS: the two mean lifetimes obtained from the statistical analysis are $\overline{\tau}_1 = 0.75$ s and $\overline{\tau} = 2.1$ s, respectively. This again confirms the model that the two types of channels are active in the two coexisting phases, each with a distinct influence on the pore characteristics.

Also the phase information derived from lifetime measurements of gramicidin in membranes of different compositions (see Figure 3b) is equivalent to the interpretation based on the conductance data: the short lived channel, indeed, corresponds to the PC–rich phase, whereas the longer lifetime is found in PG–rich domains. Again the mean lifetimes of the two populations do not change for phase separated bilayers within the miscibility gap.

One other detail is noteworthy: for homogeneous membranes the mean lifetimes of the single channels decrease considerably as one approaches a phase boundary. The (dynamical) formation of the second phase seems to be announced by a destabilization of

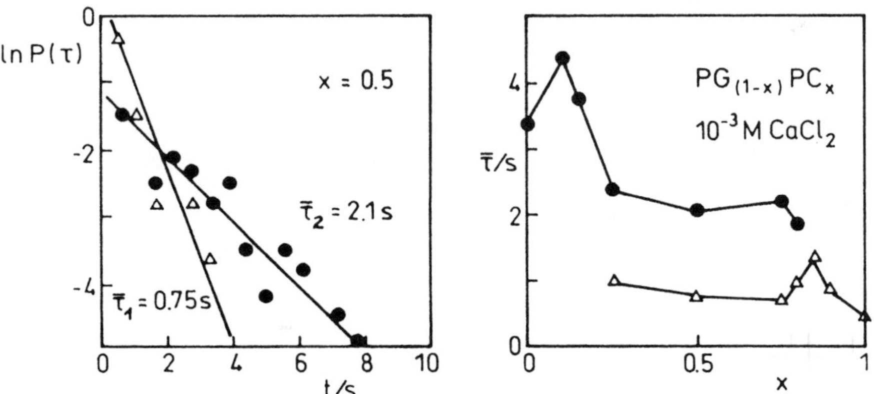

<u>Fig. 3</u>: a) Lifetime probability distribution of the two pores seen in Fig. 2a. Open triangles: pores with $\Lambda_1 = 0...40$ pS, full circles: pores with $\Lambda_2 = 40...100$ pS. b) mean lifetime, τ, of gramicidin pores in mixed or demixed membranes of various mole fractions, x, symbols like in a).

246

the gramicidin dimer—state by fluctuations in the bilayer composition. Similar phenomena had been found in BLM close to a critical demixing point (Knoll et al., 1986).

If these series of measurements are performed in the presence of different concentrations of Ca^{++} the full phase diagram can be evaluated as we will discuss in detail elsewhere (Mittler—Neher and Knoll, 1989b)). We should mention, however, that the tendency, e.g. for equimolar mixtures of PC and PG to demix can be seen already at the $10^{-5} - 10^{-6}$ M Ca^{++}—level. This offers, therefore, a very sensitive mechanism to control a membrane function by lateral structure formation.

<u>pH—dependence of phase separation</u>: In the previous section we had shown that the strong interaction of Ca^{++}—ions with the negatively charged PG headgroups caused, in some cases, a destabilization of a homogeneously mixed membrane and induced a phase separation. Since the responsible phosphate—group of PG can be titrated it should be possible — at sufficiently low pH — to turn off this demixing ability of Ca^{++}. To demonstrate this, we performed single channel experiments with 1/1 mixed PC/PG membranes in the presence of 10^{-4} M Ca^{++} at various pH—values in the CsCl electrolyte solution. The result is summarized in Figure 4: Two channel populations are found between pH 6 and pH 3.75. From pH 3 on, however, only a monomodal conductance histrogram shows a homogeneously mixed matrix. At an intermediate pH 3.5 a rather broad distribution is found.

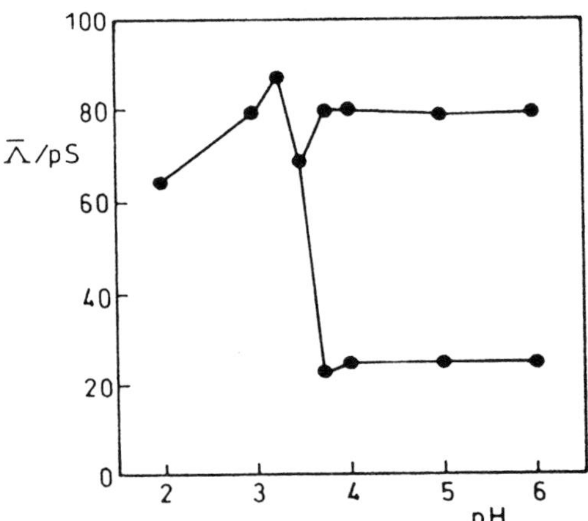

Fig. 4: Mean conductance increments, Λ, by a single gramicidin pore in a mixed or demixed membranes at different pH—values of the electrolyte (0.5 M CsCl, 10^{-4} M $CaCl_2$).

This result can be understood if one takes into account that PG changes around pH 3.5 from neutral to negatively charged (Van Dijck et al., 1978). Starting at pH 2 with a mean conductance of $\overline{\Lambda} = 65$ pS the single channel conductivity increases first as one increases the pH because a growing fraction of already dissociated PG charges the membrane negatively so that the interfacial Cs^+–concentration is enhanced. If more PG is dissociated Ca^{++}–ions come into play and eventually are sufficiently increased in their interfacial concentration that they can induce again phase separation.

Therefore, a second mechanism for order–function relation in membranes may be the (local and transient) change of the proton concentration. We should mention that details of this switching depend on the involved ions: in the presence of Zn^{++}–ions phase separation occurs already at pH 3, if Mg^{++} is employed always two channel populations over the full pH range investigated was found (Mittler–Neher, 1989).

<u>Polyelectrolyte–induced demixing</u>: As a final example for chemically driven phase separation we present results obtained with two different polyelectrolytes. The first is a positively charged polyvinylmethylpyridinium (PVMP, average molecular mass $M_r \approx 20000$) which at a concentration of 10^{-4} M induces in a 1/1 mixed PG/PC membrane demixing as can be clearly seen from the bimodal conductance histogram shown in Figure 5a). This experiment was performed at pH 6 with ultrapure CsCl solution which in the absence of PVMP, does not cause phase separation. The second example was obtained with polylysine ($M_r = 5400$, c = $5 \cdot 10^{-4}$ M) a well–known demixing poly–ion (Mittler–Neher and Knoll, 1989a), also for vesicles and liposomes (Galla and Sackmann,

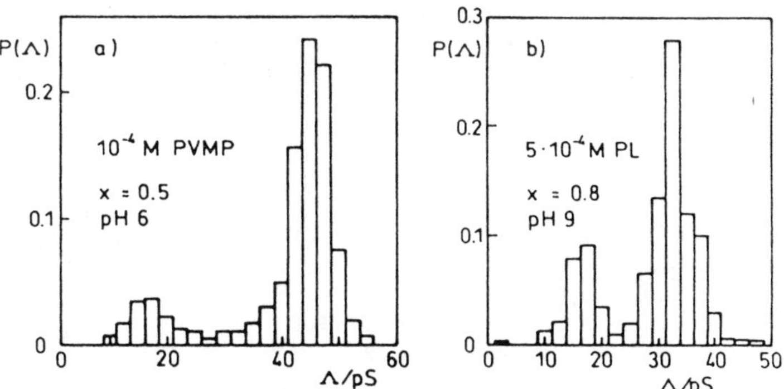

<u>Fig. 5</u>: Normalized conductance histrogram of a) a x = 0.5 membrane demixed by PVMP and b) a x = 0.8 membrane demixed by PL.

1975; Carrier and Pezolet, 1984). This time the mole fraction of PC was $x = 0.8$ and the pH adjusted to pH 9. Again two channel populations are deduced from the conductance histogram (Figure 5b). Common to both results is an obvious blocking of the gramicidin channels which seems to indicate a screening and even reloading of the charged membrane–solution interface.

CONCLUSION: The presented results fully establish the experimental possibility for studying the coupling of lipid phase separation processes to a membrane function – the transport if ions across the hydrophobic barrier by model proteins. They offer therefore a link between the well–known phase behavior of lipids and lipid alloys and many physiological processes which are controlled or at least modified by the physical state, i.e. the structure and order, of the lipid matrix into which the functional units are embedded. The framework of the Poisson–Boltzmann description of the coupling of the pH, the membrane charge density and the (bulk and interfacial) ion concentrations offers a whole variety of control mechanisms for membrane functions by lateral organization which now can be studied at the model membrane level but on sound theoretical grounds.

REFERENCES

Bamberg, E., and Läuger, P. (1973) J.Membrane Biol. 11, 177–185.
Bamberg, E., and Läuger, P. (1977) J.Membrane Biol. 35, 351–375.
Carrier, D., and Pezolet, M. (1984) Biochem. 25, 4167–4174.
Galla, H.–J., and Sackmann, E. (1975) J. Am. Chem. Soc. 97, 4114–4120.
Henkel, T., Mittler, S., Pfeifer, W., Rötzer, H., Apell, H.–J., and
    Knoll, W. (1989) Biochimie 71, 89–98.
Hladky, S.B., and Haydon, D.A. (1970) Nature 225, 451–453.
Knoll, W., Apell, H.–J., Eibl, H., and Miller, A. (1986) Eur. Biophys.J. 13, 187–193.
Mittler–Neher, S. (1989) PhD Thesis, Johannes–Gutenberg–Universität Mainz.
Mittler–Neher, S., and Knoll, W. (1989a) Biochem. Biophys. Res. Commun.,
    162, 124–129.
Mittler–Neher, S., and Knoll, W. (1989b) submitted.
Montal, M., and Mueller, P. (1972) Proc. Natl. Acad. Sci. USA 69, 3561–3566.
Mueller, P., Rudin, D.O., Tien, H.T., and Wescott, W.C..(1962) Nature 196, 979–980.
Sackmann, E. (1984a) In: Biological Membranes Vol. 5 (D. Chapman, Ed.),
    Academic Press, London, pp. 105–143.
Sackmann, E. (1984b) In: Synergetics – From Microscopic to Macroscopic Order
    (E. Frehland, Ed.), Springer Series in Synergetics, Vol. 24, Springer Verlag,
    Berlin, pp. 68–79.
Schmidt, G., Eibl, H., and Knoll, W. (1982) J. Membrane Biol. 70, 147–155.
Van Dijck, P.W.M., De Kruijff, B., Verkleij, A.J., Van Deenen, L.L.M., and
    De Gier, J. (1978) Biochim. Biophys. Acta 512, 84–96.

A STUDY OF CADMIUM AND CALCIUM TRANSPORT INTO A MARINE UNICELLULAR ALGA.

M. ROMEO (1), C.S. KAREZ (1), D. ALLEMAND (2), G. de RENZIS (3), M.GNASSIA-BARELLI (1), and S. PUISEUX-DAO (1).

(1) Unité INSERM 303, B.P.3. 06230 Villefranche-sur-mer, France.
(2) Centre Scientifique de Monaco, Musée Océanographique, 98000 Monaco.
(3) UA CNRS 651, Université de Nice, 06034 Nice-Cedex, France.

SUMMARY: ^{45}Ca and ^{109}Cd uptake were followed in *Criscosphaera elongata*. In both cases after a rapid increase for the first 5 min., the incorporation rate slowed during the hour of observation. Verapamil, a blocker of voltage dependent slow calcium channels inhibited ^{45}Ca uptake except for the first rapid phase where adsorption should predominate. Cadmium also decreased ^{45}Ca labelling suggesting antagonism between the two metals. However verapamil was shown to augment ^{109}Cd incorporation. The data support the presence of calcium channels in the alga and suggest several processes in Cd accumulation.

For the last ten years, it has been pointed out that the movement of ions, such as calcium, across the cell membrane plays a critical role in the control of cell metabolism (Carafoli, 1987; Kauss, 1987). Thus, an alteration of these movements may lead to toxic injuries (Orrenius and Nicotera, 1986; Komulainen and Bondy, 1988). Cadmium, an important pollutant to the marine environment, is known to interfere with the transport of calcium in animal cells (Hinkle *et al.*, 1987). Pick *et al.*, (1986) and Heuillet *et al.* (1988) demonstrated that cadmium may decrease the transport of calcium into phytoplankton cells. In a recent publication (Roméo *et al.*, 1989) concerning the uptake of unlabelled cadmium by *Cricosphaera (Hymenomonas) elongata*, we reported an interaction between stable Cd and Ca.

In this work, ^{45}Ca and ^{109}Cd uptake characteristics were studied in the presence and in the absence of verapamil, a blocker of voltage-dependent calcium channels. The interaction between Ca and Cd was also investigated by studying the kinetics of Ca uptake by the phytoplankton cells in the presence of Cd increasing concentrations.

MATERIALS AND METHODS

Cricosphaera elongata (Droop) Parke and Greene cells were grown in batch cultures, in natural seawater enriched with "f/2" medium (Guillard and Ryther, 1962) under sterile conditions. Cultures were kept at 18 $\pm$ 1°C under 12h light:12h dark rhythm (E = 40 W.m^{-2}). Culture growth was followed by counting cells in a Lemaur haemacytometer (triplicate counting). At the end of the exponential phase (10 days of culture), algae were concentrated to 10% of the volume by tangential ultrafiltration "Millipore" system. They were then transferred to artificial seawater medium (500 mM NaCl, 50 mM MgCl$_2$,6H$_2$O, 10 mM KCl, 2.4 mM NaHCO$_3$; pH = 8.0). Cells were kept in this low Ca medium for 48h before the uptake experiments in order to obtain significant labelling. A second ultrafiltration was carried out just before the experiments in order to withdraw almost all the culture medium and to obtain a high cell density (10^6 cells.ml^{-1}). Ca and Cd fluxes were followed with ^{45}Ca (5μCi.ml^{-1}) and ^{109}Cd (0.1 μCi.ml^{-1}). Each radio-element was added to 10 ml culture medium for various periods of incubation. Cells were incubated in the light at 20°C. The culture recipients were shaken about every two minutes during the labelling experiments. After different incubation times (from 5 min. to 1 h), 1 ml aliquots were removed and filtered through 8.0 μm "Sartorius" membrane filters. Filters were then washed three times with 5.0 ml of artificial sea water. The radioactivity in algae and in the medium were determined in a liquid scintillation counter.

RESULTS AND DISCUSSION

^{45}Ca uptake studies: The uptake of ^{45}Ca by the cells for a short period of time (60 min.) is shown in Fig.1. It can be described by a rapid phase for the first minutes followed by a slower phase. The rates of Ca uptake were 0.128 nmoles.min^{-1} per 10^6 cells for the first phase (5 min.) and 0.022 nmoles.min^{-1} per 10^6 cells for the second phase (from 5 min. to 1 h). The first phase should correspond to a dominant adsorption process at the cell envelopes while the slower phase may predominantly represent the calcium transported into the cells.

A voltage dependent calcium channel blocker, the verapamil, was used to characterize calcium uptake. This drug, a well-known inhibitor of Ca transport in animal cells. was added at the concentration of 100 μM to the culture medium and uptake experiments were carried out subsequently with ^{45}Ca. Similar concentrations are currently used for marine animal cell experiments (Komukaï *et al.*, 1985). Verapamil did not disturb the first rapid phase but inhibited ^{45}Ca influx into the cells afterwards (Fig.1). This result supports the existence of a calcium transport mediated by voltage-dependent channels and also that the first rapid uptake mainly involves adsorption and diffusion processes.

Verapamil was reported to be a calcium antagonist in sea urchin eggs at 250 μM (Komukai *et al.*, 1985). Calcium channels have been demonstrated by ^{45}Ca labelling or by electrophysiological experiments in several Characeae (algae); their activity is modified by several Ca channel inhibitors (verapamil, nifedipine) or activator (Bay K8644) (Kikuyama and

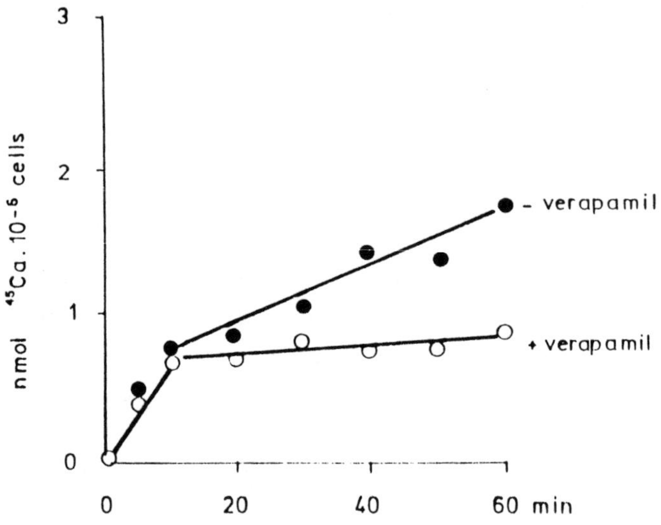

Fig. 1 : ^{45}Ca uptake by *Cricosphaera elongata*

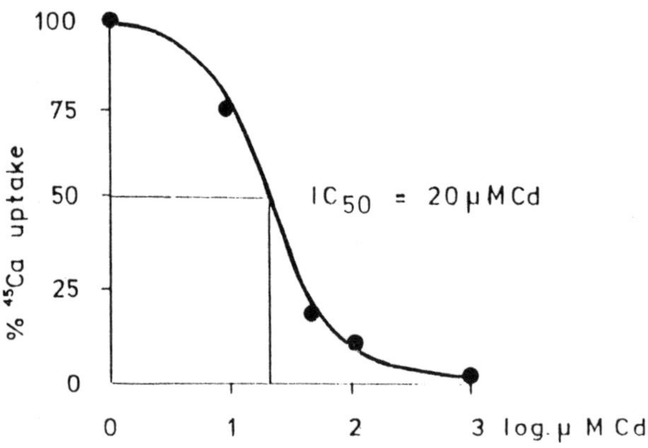

Fig. 2 Dose response curve of ^{45}Ca uptake

Tazawa, 1983; Lunevesky *et al.*, 1983; Shiina and Tazawa, 1987). Verapamil effects on several cell activities suggested their presence in the unicellular alga *Euglena* (Lonergan and Williamson, 1988) while the existence of verapamil sites was reported in another species *Chlamydomonas reinhardtii* by Dolle and Nultsch (1988).

Cd action on ^{45}Ca uptake: In this experiment, cells were exposed to ^{45}Ca in the presence of different stable cadmium concentrations for 20 min. (fig.2). An inhibition of ^{45}Ca influx into *Cricosphaera elongata* was observed as a function of Cd concentrations in the medium. The half maximal inhibition of ^{45}Ca uptake may be obtained at a Cd concentration of ca. 20 μM. This value is lower than that reported by Pick *et al.*, 1986 in the phytoplankton cell *Dunaliella salina* (IC_{50} = ca 100 μM) and may imply a higher sensitivity of *C.elongata* to cadmium.

^{109}Cd uptake studies: The kinetics of ^{109}Cd accumulation by *Cricosphaera elongata* was followed at 2.0 μM and 10.0 μM for 30 min. Fig 3 shows that, as for Ca uptake, there are two phases of absorption: a rapid uptake of ^{109}Cd for the first 5 min., followed by a slower accumulation.

Cricosphaera elongata was incubated for 5 min. (i.e. by the end of the first phase of uptake) with Cd concentrations ranging from 2 nM to 20 μM. The uptake by the cells (expressed as log) increased linearly with the logarithm of the metal concentration in the medium, suggesting, as for calcium, that the first phase of cadmium uptake may be interpreted as a dominant adsorption by binding to the cellular envelopes. Rebhun and Ben-Amotz (1984) proposed that the marine microalga *Chlorella stigmatophora* can accumulate Cd by adsorption on the cell wall and/or by diffusion into the cells against a concentration gradient, or by a process requiring energy.

The effect of verapamil (100 μM) was tested on cumulative Cd

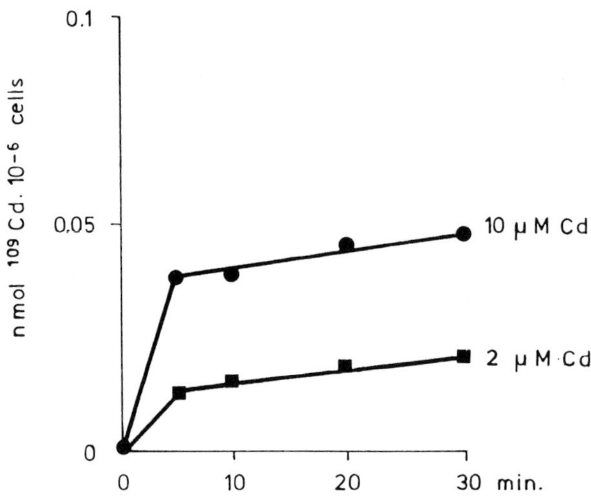

Fig. 3 : 109Cd uptake by *Cricosphaera elongata*

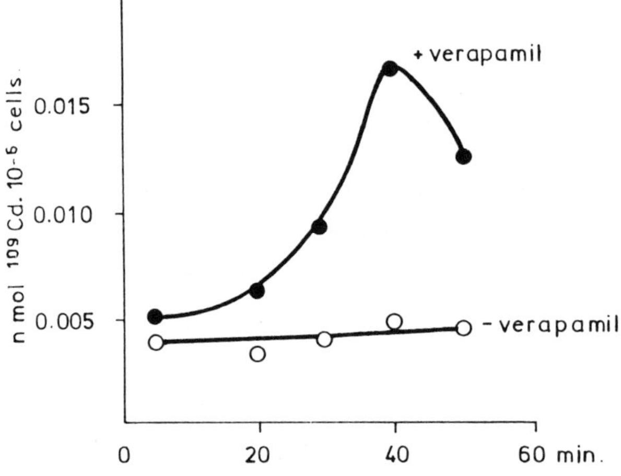

Fig. 4 : Effect of verapamil on 109Cd uptake (pulse labelling).

uptake (2 μM in the medium); a high stimulation of this uptake was observed with a maximum after ca 50 min. incubation. The same experiment was carried out in which a 2 min. pulse [109]Cd labelling was performed every ten minutes after the introduction of verapamil. Fig.4 shows that verapamil drastically increased Cd uptake. The time course of verapamil action displayed a similar pattern than that observed in cumulative uptake experiments. Following the maximum observed after ca 40 min. of incubation with verapamil, Cd influx displayed a progressive decrease. Fig.4 also shows that the constant radioactive level observed after 2 min. labelling in the absence of verapamil likely indicated the binding to cellular envelopes. Then verapamil seems to stimulate a process that does not work in control. To explore the link between Ca and Cd, we measured in 2 min. pulse experiments, the verapamil-stimulated [109]Cd influx at two concentrations of external stable Ca. For example, the effect of verapamil measured 60 min. after addition of this compound was shown to depend on the concentration of calcium (1 mM $CaCl_2$ outside, [109]Cd influx: 12 $pmole.min^{-1}.10^{-6}$ cells; 10 mM $CaCl_2$ outside, [109]Cd influx: 5 $pmole.min^{-1}.10^{-6}$ cells). These results suggest the existence of a competition between the two ions. Contrary to what happens in animal cells (Hinkle *et al.*, 1987) verapamil stimulated Cd uptake by *Cricosphaera elongata*. An explanation could be that a high percentage of dissolved Cd in sea water is under the neutral chemical form $CdCl_2$, the lipophilic drug could promote $CdCl_2$ diffusion through the membrane either directly or indirectly. Once inside the phytoplankton cells, Cd could compete with calcium taking advantage of the inhibition of Ca uptake by the drug.

CONCLUSIONS
____

The action of verapamil on [45]Ca uptake supports the existence of voltage-dependent calcium channels in the cell membrane of *Cricosphaera elongata*. High Cd concentrations inhibit cumulative

^{45}Ca uptake by the cells, the half inhibition concentration reaches 20 μM Cd after 20 minutes.

Contrary to what occurs in animal cells, verapamil enhances Cd influx which could be due to Cd speciation in seawater. The increase of Cd uptake depends on Ca concentration in the medium which reflects Ca-Cd competition such as Cd inhibition of ^{45}Ca incorporation into *Cricosphaera elongata*.

Acknowledgments: This research was partly supported by Centre National de la Recherche Scientifique CNRS AIP 06931-1988 and by a Franco-Brazilian program CEFI-CNPq: Comité de Formation d'Ingénieurs-Conselho Nacional de Desenvolvimento cientifico et Tecnologico

REFERENCES

Carafoli, E. (1987) Ann. Rev. Biochem., 56, 395-433.
Dolle, R., and Nultsch, W. (1988) Arch. Microbiol., 149, 451-458.
Guillard, R.R.L., and Ryther, J.H. (1962) Can. J. Microbiol., 8, 119-239.
Heuillet, E., Bermond, A., Jeanne, N., Puiseux-Dao, S., and Ducauze, C. (1988) Mar. Ecol. Prog. Ser., 44, 69-75.
Hinkle, P.M., Kinsella, P.A., and Osterhoudt, K.C. (1987) J. Biol. Chem., 262, 16333-16337.
Kauss, H. (1987) Ann. Rev. Plant Physiol., 38, 47-72.
Kikuyama, M., and Tazawa, M. (1983) Protoplasma, 117, 62-67.
Komukai, M., Fujiwara, A., Fujino, Y., and Yasumasu, I. (1985) Exp. Cell. Res., 159, 463-472.
Komulainen, H., and Bondy, S.C. (1988) TIPS, 91, 154-156.
Lonergan, T.A., and Williamson, L.C. (1988) J. Cell Sci., 89, 365-371.
Lunevesky, V.Z., Zherelova, O.M., Vestrikov, I.Y., and Berestorky, G.N. (1983) J. Membrane Biol., 72, 43-58.
Orrenius, S., and Nicotera, P. (1986) Klin. Wochensch., 64 (suppl. VIII), 138-141.
Pick, U., Ben-Amotz, A., Karni, L., Seebergts, C.J., and Avron, M. (1986) Plant Physiol., 81, 875-881.
Rebhun, S., and Ben-Amotz, A. (1984) Water Res., 18, 173-179.
Roméo, M., Gnassia-Barelli, M., Karez, C.S., and Puiseux-Dao, S. (1989) Toxicol. and Environm. Chem., 22, 97-100.
Shiina, T., and Tazawa, M. (1987) J. Membrane Biol., 96, 263-276.

INVOLVEMENT OF D_2O FOR H_2O SUBSTITUTION
IN THE BEHAVIOUR OF SOME EXCITABLE MEMBRANES

V. Vasilescu, Cornelia Zaciu and Mioara Florica Tripşa

Department of Biophysics, Faculty of Medicine, 76241 Bucharest, Romania

ABSTRACT: a) Heavy water significantly modifies the ionic currents both in their amplitude and in their time course; b) It is shown in our experimental studies concerning the quasitotal deuteration that the molecular mechanisms reflected at macroscopic level by the ionic currents and the excitation threshold of the nerve membranes are strongly affected when the proton-deuteron exchange is produced in the intracellular medium; c) Taking also into account the data concerning the ouabain effects on the excitation threshold of the node of Ranvier of the frog sciatic nerve fiber as well as those concerning decreasing the ATP pool of the deuterated frog sciatic nerve under electrical stimulation, we can plead for an excitation-energy coupling which is more evident at high frequency of stimulation.

INTRODUCTION

The basic idea of our previuos and recent studies is that the replacement of water by heavy water is a useful nondestructive method capable of revealing the participation of water and particularly of protons at various biological processes, including energetics. (Ahmed & Foster, 1974; Kols, 1975; Vasilescu et al., 1976; Lobyshev et al., 1978; Vasilescu & Zaciu, 1980, 1983; Zaciu & Vasilescu, 1981; Vasilescu et al., 1987; Tripşa et al., 1988).

The investigations performed in our Laboratory concerning the effects of deuteration on the bioelectrogenesis of some biological systems which have specialized excitable structures - frog heart and retina - showed a number of phenomena among which the most interesting that we consider are the following:

- a dramatic change of the latency of the OFF response of retinogram which increases from 50 to 200 msec, while the latency of the ON response remains almost unmodified (Chirieri et al., 1977);

- the blockage of the mechanical activity of the heart 10 minutes after perfusion with D_2O-Ringer and the blockage of the electrical activity of the heart 30 minutes after similar perfusion; both activities continue more than 2 hours when H_2O-Ringer instead of D_2O-Ringer perfusion is used (Vasilescu & Ciureş, 1978).

Taking into account these results as well as the data reported by others (Meves, 1974; Schauf and Bulloch, 1979, 1980; Schauf, 1983) we studied furthermore the effects of deuteration on the ionic currents and on the dynamical behaviuor of the node of Ranvier membrane from the frog sciatic nerve fiber.

D_2O action on some energetical processes have been revealed from experiments following its effects on ATP pool of stimulated frog sciatic nerve and from comparative study between ouabain and D_2O effects on the behaviour of the node of Ranvier membrane.

Our results are discussed in terms of possible molecular mechanisms involved in the excitation processes of the nerve membrane which were revealed by deuteration.

MATERIALS AND METHODS

Experiments were carried out on the myelinated fibers from the small motor branch of the frog sciatic nerve (Rana esculenta). The dissection and setting up the sample in the chamber have been performed after Stampfli's method (Stampfli, 1969); a special microdissection table with dark field illumination has been used.

Ionic currents (see the first subsection of RESULTS) were recorded with a voltage-clamp device according to Nonner's method (Nonner, 1969; Zaciu et al., 1981).

The double air gap arrangement, previously reported (Vasilescu and Zaciu, 1980) was prefered for the dynamical behaviour studies; it allowed placing the stimulus on one internode adjacent to the node under investigation and recording the action potential on the other internode.

The external deuteration was performed by maintaining a continuous D_2O-Ringer flow in the compartment of the node to be investigated. Total deuteration was achieved by introducing D_2O-Ringer since the last stage of the dissection and afterwards by immersing the fiber 3-4 times in D_2O-Ringer; all compartments of the chamber where the fiber has had to be placed were filled with D_2O-Ringer, too.

Solutions: The following composition for Ringer solution (both in H_2O and D_2O) has been used; **NaCl** (110mM/l); **KCl** (2.5mM/l); **CaCl$_2$** (1.8 mM/l); Tris-Cl (5 mM/l): **pH** and **pD** values were adjusted at 7.4 and 7 respectively. A Beckmann (PHASAR-1) **pH**-meter with glass electrode was employed to measure **pH** and **pD**.

The following solution has been used in the experiments concerning the ouabain effects: **NaCl** (110 mM/l); **KCl** (2.5 mM/l); **CaCl$_2$** (1.8 mM/l); ouabain (10^{-3}) mM/l); **Tris-Cl** (5 mM/l) (**pH** = 7.4).

RESULTS

Deuteration effects on the ionic currents: The most important results that we have obtained in our studies concerning the deuteration effects on the ionic currents of the node of Ranvier membrane are the following:

a) The amplitude of the ionic currents drastically decreases (see Fig. 1 and Fig. 2) under quasitotal deuteration (t = 18°C).

b) The amplitude and the time course of the ionic currents are almost the same as in **D$_2$O** (Fig. 3) when only the node under investigation is reimmersed in **H$_2$O**-Ringer solution, the rest of the fiber being kept in **D$_2$O**-Ringer.

c) A spectacular recovery of the ionic currents (both in their amplitude and in their time course) is seen when the whole fiber is reimmersed in **H$_2$O**-Ringer solution (total-rehydration) (Fig. 4).

d) The selectivity of the sodium channels calculated from the reversal potential measurements was not significantly affected by the deuteration.

Thus, the selectivity ratio given by formula:

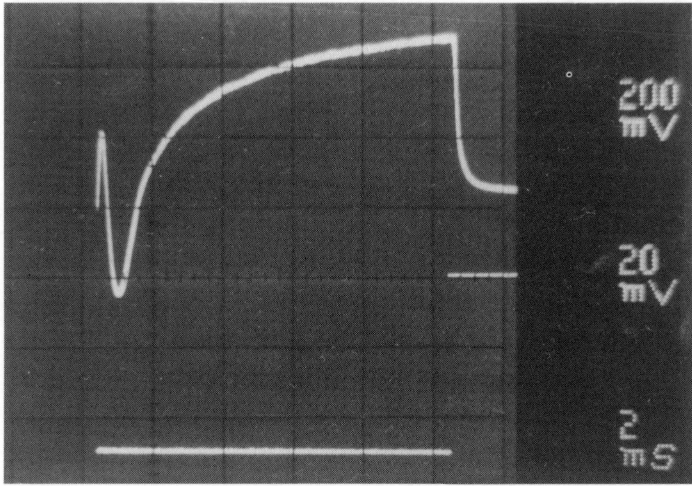

Fig. 1. Ionic currents of the quasi-totally deuterated fiber recorded after 1-3 minutes of stimulation.

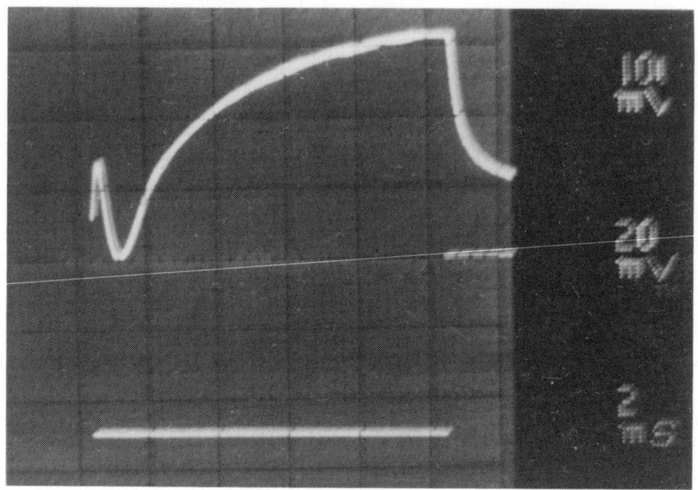

Fig. 2. Ionic currents of the quasi-totally deuterated fiber recorded after 15 minutes of stimulation.

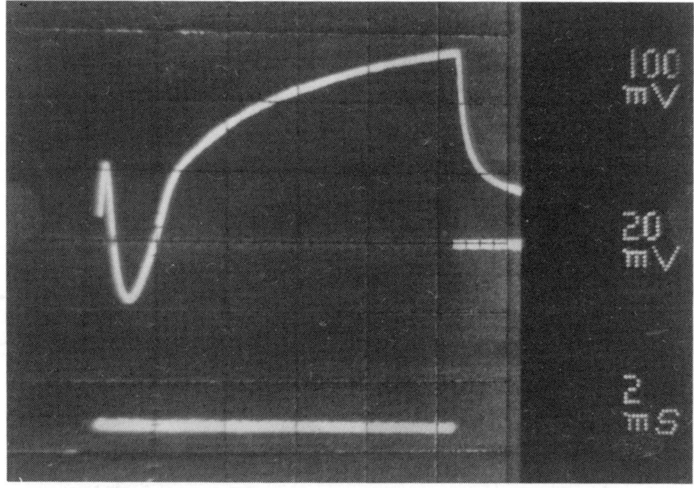

Fig. 3. Ionic currents of the fiber recorded after external rehydration (when only the node to be investigated is reimmersed in H_2O-Ringer).

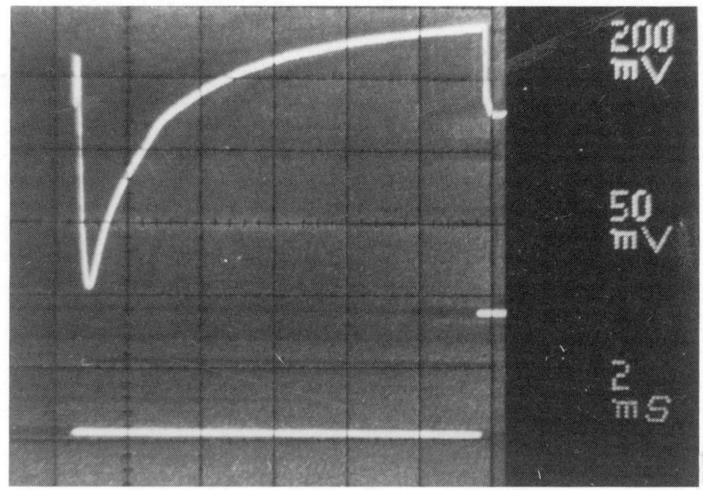

Fig. 4. Ionic currents of the fiber recorded after total rehydration (when the whole fiber is reimmersed in H_2O-Ringer).

$$S = \frac{(P_{Na}/P_K)_{H_2O}}{(P_{Na}/P_K)_{D_2O}} = \frac{\exp\frac{F}{RT} V_{reversal\ H_2O}}{\exp\frac{F}{RT} V_{reversal\ D_2O}} \tag{1}$$

has an average value of 1.2 (see Table I and Table II).

Table I. Reversal potentials for the ionic currents of the node of Ranvier membrane as resulted from voltage clamp measurements.

Fiber	$V_{reversal\ H_2O}$ (mV)	$V_{reversal\ D_2O}$ (mV)	$\Delta V_{reversal}$ (mV)
1	+51.5	+46	+5.5
2	+51	+47	+4
3	+55	+44	+6
4	+57	+52	+5
5	+52	+48	+4
6	+53	+50	+8

Table II. Selectivity ratios calculated from the reversal potentials

Fiber	$F \Delta V_{reversal}/RT$	$S = \dfrac{(P_{Na}/P_K)_{H_2O}}{(P_{Na}/P_K)_{D_2O}}$
1	219.145×10^{-3}	1.245
2	159.38×10^{-3}	1.173
3	239.07×10^{-3}	1.270
4	199.255×10^{-3}	1.220
5	159.38×10^{-3}	1.173
6	318.76×10^{-3}	1.375

In the above formula P_{Na}/P_K is the permeability ratio for sodium and potassium ions, F, R, T have their termodynamic meaning and $V_{reversal}$ is the potential for which the transmembrane current is zero. It should be point out that the same value that we found for S has been reported in other experimental studies concerning ionic mobilities in heavy water for water substitution (Garby and Nordquist, 1955).

D_2O effects on the dynamical behaviour of the node of Ranvier membrane: By the dynamical behaviour we mean the way in which the membranes function during rhytmic stimulation and when as a result, the action potential appears and can be recorded.

Though the dynamical behaviour involves complex phenomena, we decided to investigate the excitability under these conditions – natural ones, under which an excitable membrane functions – by using deuteration and high-frequency stimulation (over 100 Hz) in order to reveal how the proton-deuteron exchange could influence certain emphasizing mechanisms in the node of Ranvier membrane.

The variation of the excitation threshold as function of the stimulation frequency under different conditions of the immersion of the fiber is shown in Fig. 5.

Our measurements point out the fact that the maximum stimulation frequency at the constant threshold has a value of 50 Hz when the short-term pulses (0-100 μs) are used; the threshold increases by 30% at 100 Hz, while at frequencies of more than 200 Hz an increment of almost 80% of the threshold if found.

The excitability of the fiber under quasitotal deuteration have evolved as follows:

a) The excitability is lost after 15 minutes of continuous stimulation; the nerve fiber exposed to the similar quasitotal deuteration recovers its excitability when return to a low frequency of stimulation.

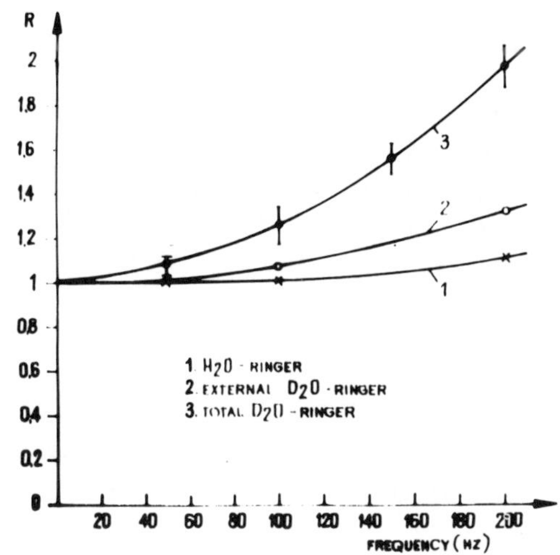

Fig. 5. Variation of the excitation threshold as function of the stimulation frequency under different conditions of fiber immersion; curves (1) and (2) are plotted for the same membrane; curve (3) is a plot of averages of 10 measurements, each R representing the ratio of the threshold at any frequency f to that at the frequency of 10 Hz.

b) The rehydration effects after the loss of the excitability induced by quasitotal deuteration and high frequency stimulation are as follows:

- the excitability continues to be inhibited during stimulation at high frequencies and external hydration, when only the node under investigation is immersed in H_2O-Ringer, all other compartments being maintained in D_2O-Ringer;

- the excitability reappears at high frequency of stimulation during total hydration, when the whole fiber is kept in H_2O-Ringer, the phenomenon having a rapid kinetics.

How can we explain the results concerning the quasitotal deuteration effects on the dynamical behaviour of the node of Ranvier membrane? A possible hypothesis is the following: reappearance of the excitability at low frequency of stimulation after it was lost by quasitotal deuteration and high frequency of stimulation of the nerve fiber strongly suggests the existence of one or more mechanisms coupled with the excitability processes at high frequency of stimulation.

The abolishment of the excitability induced by quasitotal deuteration and rhythmic stimulation becomes reversible after total reimmersion in H_2O-Ringer; therefore, the mechanism(s) mentioned above is (are) drastically affected by proton-deuteron exchange in the intracellular medium.

Taking also into account the data in which a substantial decrease of ATP pool of the deuterated nerve during stimulation is shown to occur (Vasilescu et al., 1976) we may conclude that our results are consistent with idea of an excitation-energy coupling which is more evident at high frequency of stimulation.

Our results are also in good agreement with those obtained by Maximov and his coworkers concerning the stimulation effects on the ATP pool and **Na-K** ATPase activity of the nerves from frog, rat and lobster (Maximov et al., 1978). They revealed pronunced changes of both the ATP pool and ATPase activity, the maximum being found at 10 Hz for the frog sciatic nerve.

Comparative studies between ouabain and deuteration effects on the behaviour of the node of Ranvier: Taking into account the results presented in the previous subsection we decided to perform some experiments concerning the ouabain effects on the excitability of the node of Ranvier membrane. Our main interest was focused again on the excitation threshold.

Our intention was to block the sodium pump by this specific inhibitor and to see how this is reflected in the excitability processes. These kind of experiments, in which a concentration of 10^{-6}M of ouabain has been used, revealed that the maximum frequency of stimulation for constant threshold decreased to half after 15 minutes of long-term stimulation (Fig. 6).

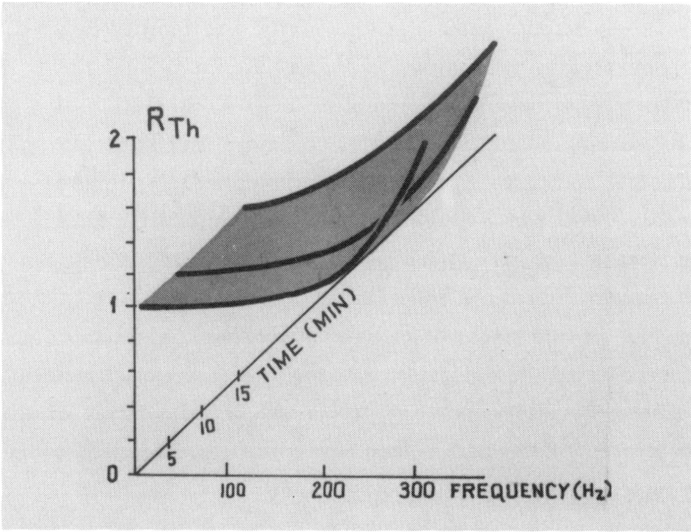

Fig. 6. Variation of the treshold (R_{Th}) versus frequency and time for the node of Ranvier immersed in ouabain-Ringer solution 10^{-6}M.

The effects of external deuteration are characterised by a rapid kinetics, while those of ouabain (10^{-6}M) appear relatively slow.

The similarities between the effects of ouabain and the external deuteration suggest that the inhibition of the sodium pump activity could be at least one of the possible mechanisms of the heavy water action; that can be done by preventing the occurrence of the enzyme conformation changes during the sodium dependent phosphorilation step and/or during the stage of coupling the potassium at the dephosphorilation step (Fig. 7 and Fig. 8).

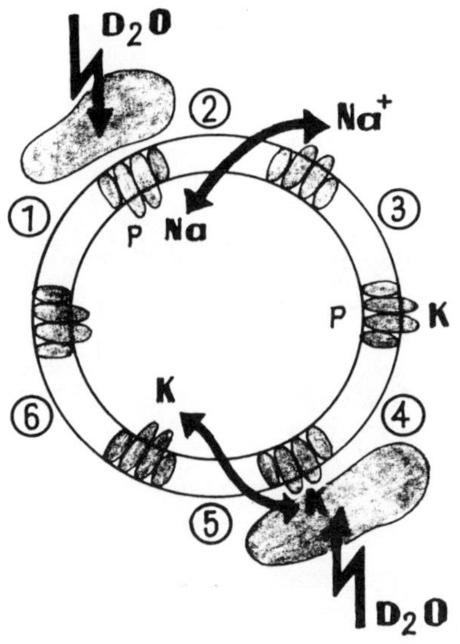

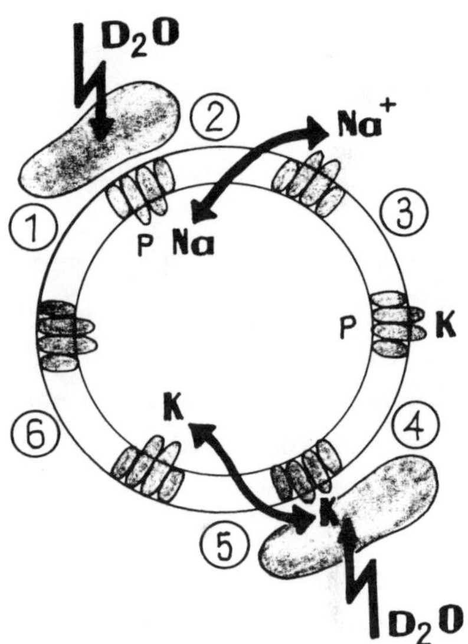

Fig. 7. Simplified model illustrating the steps of the sodium pump conformational changes and the site of ouabain action.

Fig. 8. Simplified model illustrating the steps of the sodium pump conformational changes and the possible sites of action of the heavy water.

The steps mentioned above are corresponding to the following biochemical reactions, in which H_2O could be exchanged by D_2O:

1. $E_1 + ATP \rightleftharpoons E_1(ATP)$

$$E_1(ATP) + 3Na^+ \xrightleftharpoons{Mg + H_2O} E_1 \sim P(Na^+)_3 + ADP$$

2. $E_1 - P(Na^+)_3 + Mg^{++} \rightleftharpoons E_2 - P + 3Na^+$

3. $E_2 - P + 2K^+ \rightleftharpoons E_2 - P(K^+)_2$

$$4.\ E_2 - P(K^+)_2 + H_2O \rightleftharpoons E_2(K^+)_2 + P$$

$$5.\ E_2(K^+)_2 \rightleftharpoons E_1(K^+)_2 + Mg^{++}$$

$$6.\ E_1(K^+) \rightleftharpoons E_1 + 2K^+$$

Landowne's studies concerning the effects of D_2O on the sodium pump showed that the heavy water could be competitive with potassium coupling (Landowne, 1987).

The effects that we observe in our experiments are more pronounced at high frequency of stimulation because in this case the sodium pump is much more required for recovery of the ionic equilibrium.

CONCLUSIONS

a) D_2O significantly modifies the ionic currents, both in their amplitude and in their time course.

b) According to our experimental data concerning the quasitotal deuteration, the molecular mechanism(s) responsible for drastical effects of D_2O, reflected in the ionic currents (see the first subsection of RESULTS) and in the excitation threshold (see the second subsection of RESULTS) at macroscopic level of events are strongly affected by the proton-deuteron exchange in the intracellular medium.

c) Based on the experimental data concerning the ouabain effects on the excitation threshold of the node of Ranvier as well as data concerning the substantial decrease of ATP pool of the deuterated nerve during stimulation, we may conclude that our results could be consistent with idea of excitation-energy coupling; this coupling is more evident at high frequency of stimulation.

REFERENCES

Ahmed, K., and Forster, D. (1974) Ann. N.Y. Acad. Sci. **242**, 280-292.

Chirieri, E., Aricescu, I., Ganea, C., and Vasilescu, V. (1977) Naturwiss. **66**, 149.

Garby, L., and Nordquist, P. (1955) Acta Physiol. Scand. **34**, 162.

Kols, O.R., Maksimov, V.G., and Fedorov, G.E. (1975) J. Physiol. Physiological J. URSS (Russian), **LXV**, 678-685.

Landowne, D. (1987) J. Membrane Biol. **96**, 277-281.

Lobyshev, V.I., Tverdislov, V.A., Vogel, J., and Yakovenko, V.L. (1978) Biophyzika (Russian), **XXIII**, 390-391.

Maximov, V.G., Fedorov, G.E., and Kols, R.O. (1978) Biophyzika (Russian), **XXIII**, 553-559.

Meves, H. (1974) J. Physiol. **243**, 847-867.

Nonner, W. (1969) Pflugers Arch. Eur. J. Physiol. **309**, 176-192.

Schauf, L.C., and **Bulloch, J.O.** (1979) Biophys. J. **27**, 193-208.

Schauf, L.C., and **Bulloch, J.O.** (1980) Biophys. J. **30**, 295-306.

Schauf, L.C. (1983) J. Physiol. **337**, 173-182.

Stampfli, R. (1969) In: Laboratory Techniques in Membranar Biophysics (H. Passow and R. Stampfli, Eds), Springer-Verlag, Berlin, pp. 158-166.

Tripşa, F.M., **Zaciu, C.,** and **Vasilescu, V.** (1988) In: Water and Ions in Biological Systems (P. Lauger, L. Packer, and V. Vasilescu, Eds), Birkhäuser-Verlag, Basel--Boston-Berlin, pp. 173-183.

Vasilescu, V., and **Ciureş, A.** (1978), Studia Biopys. **71**, 81-84.

Vasilescu, V., **Zaciu, C.,** and **Truţia, E.** (1976), Rev. Roum. Physiol. **13**, 35-38.

Vasilescu, V., and **Zaciu, C.** (1980) Studia Biophys. **78**, 107-118.

Vasilescu, V., and **Zaciu, C.** (1983), Rev. Roum. Morphol. Embryol. Physiol. **20**, 85-88.

Vasilescu, V., **Zaciu, C.,** and **Tripşa, M.** (1987) Rev. Roum. Morphol. Embryol. Physiol. **24**, 139-143.

Zaciu, C., and **Vasilescu, V.** (1981) Studia Biophys. **84**, 71-72.

Zaciu, C., **Tripşa, M.,** and **Vasilescu, V.** (1981) Biophys. J. **36**, 757-802.

MISCELLANEOUS

FEMTOSECOND DYNAMICS OF SINGLE ELECTRON TRANSFER IN AQUEOUS MEDIA AND MIMETIC MODELS OF BIOAGGREGATES

Y. Gauduel, S. Pommeret, A. Migus, N. Yamada, and A. Antonetti

Laboratoire d'Optique Appliquée, INSERM U275,
Ecole Polytechnique-ENS Techniques Avancées, 91120 Palaiseau, France.

SUMMARY: The primary events of one-electron transfer reactions in ionic aqueous solutions and organized assemblies containing bio-molecules are investigated at the femtosecond time scale using ultrahsort UV laser pulses (100 fs duration, 1 fs = 10 $^{-15}$ s. Our results show the interest of femtosecond photochemical investi-gations in the understanding of primary steps of bioradical reactions involving ultrafast electron transfer in condensed media.

INTRODUCTION

The role of modern biochemistry is to understand how chemical reactions take place through a series of elementary steps which can occur very fast. The ultrafast reactivities of electrons in aqueous media are of fundamental importance in biology and bio-physics because the polar nature of the solvent and the energetics or time-dependence of the electron-water interactions can impede or assiste redox reactions. Electron or derived radicals can be used as a microprobe of the local structure and for the dynamics of the molecular reorganization induced by the presence of a local field or charge. The general discusion has been continuing over the relative role of the solvent friction (collisional and die-lectric components) in the frequency factor for the reaction (Razem and Hamill, 1977; Maroncelli et al., 1988, Williams, 1989).

The recent advances in the generation and measurement of ultrashort laser pulses has allowed to push investigations in the field of redox reaction dynamics down to the picosecond and even femtosecond time scale (Jonah et al., 1977, Gauduel et al., 1988[a], 1989). In the following sections we will present the results of recent investigations performed on primary events of single electron transfer reactions occuring in aqueous media and organized bioaggregates.

MATERIALS AND METHODS

<u>Samples</u>: Aqueous solutions and organized assemblies (ionic aqueous micelles and reversed micelles) are prepared with bideionized and doubly distilled water. Surfactants (NaLS, CTAB and AOT) from Sigma and Merck are of purist quality.

<u>Femtosecond laser spectroscopy</u>: The topic of generating subpico-second pulses has been discussed at length in previous paper (Migus et al., 1985, Gauduel et al., 1988[b]).

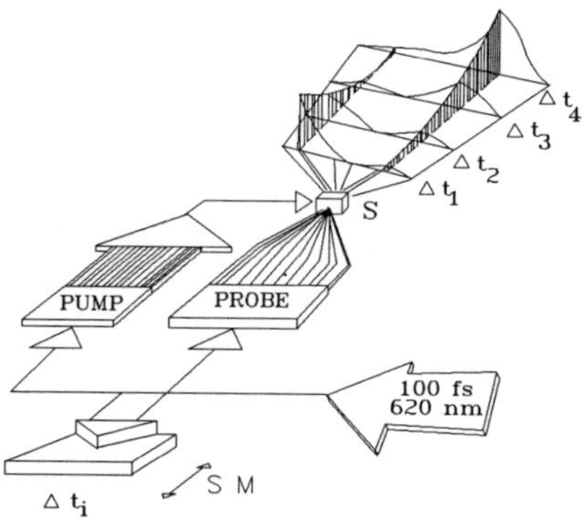

FIGURE 1. Scheme of time-resolved spectroscopy using femtosecond amplified pulses (S: sample, SM: stepping motor).

The specific points used for femtosecond investigations of single electron transfer in biomolecular systems are presented in this section. The generation of high power femtosecond laser is based on colliding pulse mode-locked dye laser (CPM) containing two following jests: one jet of rhodamine 6G and one very thin jet of saturable absorber. The pulses generated in this ring cavity are as shorter as $80 \cdot 10^{-15}$ s and the wavelength is centered around 620 nm. The amplification chain allows outup pulses of energy above 1 mJ and typically 100 fs duration at a 10 Hz repetition rate. The obtaining of high intensity femtosecond pulses are fundamental for spectroscopic purpose (figure 1) in that they can generate white light pulse of comparable duration (so called "continuum"). This continuum is used as a weak delayed pulse beam while the intense pulse, required for perturbing the analyzed system remains in the initial fundamental pulse or with less energy in its second harmonic using a thick KDP crystal ($20\mu J$ at 310 nm). This ultraviolet light constitutes the pump pulse used to initiate photochemical reactions i.e. photoionization of electron donor.

<u>RESULTS</u>

Dynamics of electron-water molecules interactions.

Photochemical studies using ultraviolet femtosecond pulses made possible new advances in the elucidation of detailed mechanisms of electron reactivity in aqueous solutions containing ions or biomolecules. The table I summarizes the values of transient concentration of hydrated electron obtained in homogeneous aqueous solutions and organized assemblies, two picoseconds after the femtosecond energy deposition in the sample. It is interesting to notice that in organized assemblies, the ionization cross-section and electron escape probability are not influenced by the surface electric potential (ψ_m). In aqueous microdroplets of reversed micelles, it can be observed a dependence of the electron capture and hydration efficiency on water pool size defined as $W = [H_2O]/[AOT]$.

In pure liquid water and organized assemblies, the photoionization of water molecules or chromophore (phenothiazine) by near femtosecond U.V. pulses has permitted to investigate some of the primary events occuring from the photoejection of epithermal excess electron (E<3eV) to the appearance of the fully relaxed hydrated electron (e^-_{hyd} or e^-_s). The excess electron (e^-_{qf}) is temporally localized in the polar medium (trapped electron). This unstable species, infrared absorbing (e^-_{ir}), fully relaxes towards a lower energy state in which water molecules are reorganized in an equilibrium configuration (e^-_{hyd}). The times of electron trapping (t_1) and hydration (t_2) are compiled in the table I.

TABLE I: Transient concentration of hydrated electron obtained 2 ps after the photoionization step and dynamics of electron trapping (localization) and hydration in homogeneous aqueous solutions or organized assemblies.

Media	$[e^-_{hyd}]$ 2 ps	e^- trap t_1 (fs)	e^- hyd t_2 (fs)
H_2O	3 μM	110 +/- 20	240 +/- 20
NaLS(0.1M) micelles	17 μM	250 +/- 20	270 +/- 30
CTAB(0.1M) micelles	17 μM	250 +/- 20	270 +/- 30
Reversed micelles			
(W=5)	1.8 μM	-	-
(W=50)	5.8 μM	140 +/- 20	270 +/- 20

The rate of formation of the trapped electron (e^-_{ir}) is lengthened in the aqueous phase of ionic micelles. The difference with pure water can be linked to the existence of an interfacial layer which modifiates the thermalization distance of the photoelectron and the primary energy loss rate of epithermal electron before trapping.

In pure water, the figure 2 shows the primary electron transfer events occuring after the photoejection of an epithermal excess electron. An early decay of the fully relaxed hydrated electron has been observed in the first 100 ps following the charge separation ($e^-..H_2O^+$). This decay is assigned to the geminate pair recombination of $(H_3O^+..e^-)_s$ or $(OH...e^-)_s$.

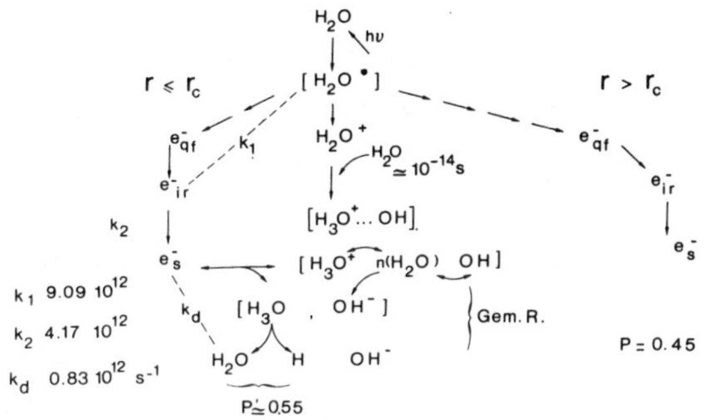

FIGURE 2: Primary species and events occuring in pure liquid water following photoejection of an epithermal electron (E<2 eV).

A fraction (55%) of the excess electrons population which have trapping and hydration distance smaller than the Onsager radius (r_c) undergoes a fast geminate recombination process because the long-range coulomb type interactions are not negligible. A significant isotope effect has been recently obtained (in preparation). The analytical solution of a geminate recombination controlled by 1D diffusion [$N_{(t)} = N_i \; erf[T_d/t]^{1/2}$] yields $T_d = 1.2$ +/- 0.1 ps for H_2O and 2.5 +/- 0.2 ps for D_2O (figure 3).

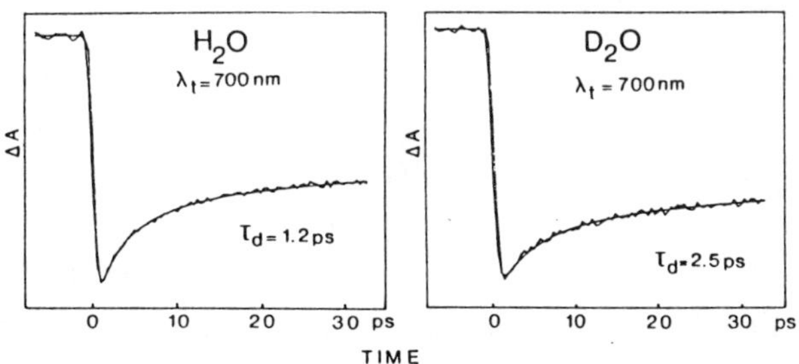

FIGURE 3: Time-resolved geminate recombination following femtosecond UV excitation (λ=310 nm) of H_2O and D_2O. The smooth lines represent the theoretical best fits of the data with a jump rate of 0.83 10^{12} s^{-1} for H_2O and 0.42 10^{12} s^{-1} for D_2O.

Ultrafast reactivity of electron with pyridine nucleotide.

The incorporation of oxidized nucleotides (βNAD$^+$) in the aqueous phase of anionic micelles (NaLS 0.07M) leads to a synthetic redox system in which phenothiazine (PTH) is the electron donor.

Table II: One electron transfer in biomicelle containing βNAD$^+$.

Species (t=2 ps)	Concentration of NAD+ (mM)	
	0	100
[e$^-_{hyd}$]	17.5 μM	14.5 μM
[NAD$^\circ$]	0 μM	2.1 μM

The table II shows the transient concentration of e$^-_{hyd}$ and pyridinyl radical (NAD$^\circ$) obtained 2 ps after the femtosecond ionization of PTH in function of the coenzyme concentration. The presence of NAD$^+$ reduces the initial e$^-_{hyd}$ concentration by 18%. Concomitantly, an early population of pyridinyl radical (2.1 μM) can be observed at 2 ps. Additional observations permit to differenciate two channels in the one electron transfer reaction (table III).

Table III: Time dependence of univalent reduction of βNAD$^+$ in micellar systems (PTH/NaLS/NAD$^+$, pH 2.7). The theoretical concentration of NAD$^\circ$ are obtained from bimolecular reaction mechanism.

Time	Observed [NAD$^\circ$]	Theoretical [NAD$^\circ$]
2 ps	2.1 +/- 0.2 μM	0.15 μM
200 ps	10.8 +/- 0.5 μM	11.1 μM

For t>200 ps, the reaction (NAD$^+$ + e$^-_{hyd}$ $\rightarrow$ NAD$^\circ$) corresponds to a bimolecular process. The rate constant k(e$^-_{hyd}$ + NAD$^+$) is pH dependent: 2.6 10^{10} M^{-1} s^{-1} (pH 6.8) and 6.8 10^{10} M^{-1} s^{-1} (pH 2.7). Bimolecular kinetics analysis with Noyes' equations demonstrate a concentration dependence of the NAD$^+$ reduction rate. The transient spectral data (figure 4) summarize the early reactions occuring in this redox system. The formation of the pyridinyl radical would correspond to a monoelectronic transfer occuring when e^-_{hyd} and NAD^+ are in physical contact i.e. when the radius of the reaction equals the sum of the e^-_{hyd} and NAD^+ radii (2.5 and 2.8 angströms respectively at pH 6.8).

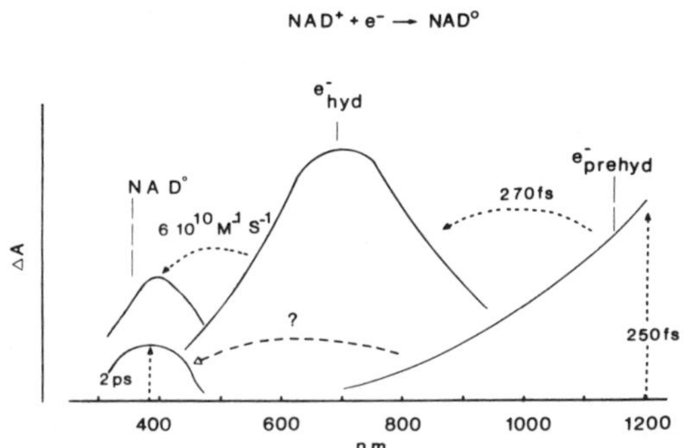

FIGURE 4: Synthetic spectral representation of primary species identified in biomolecular system (PTH/NaLS/NAD$^+$).

At shorter times (t=2 ps), the yield of the early NAD^+ radical (2.1 μM) agrees with the initial reduction of e^-_{hyd} concentration as shown in the table II. This ultrafast free-radical reaction coupled to a one electron transfer cannot be interpreted by a bimolecular mechanism because at 2 ps, only 7 % of the early NAD° population can be produced through this mechanism (table III).

<u>DISCUSSION</u>

In aqueous media, the femtosecond investigation of one electron transfer reactions involving components of living cells (interfacial water, oxidized coenzymes) permit the observation of primary redox steps. The identification of a transient state of electron in water (e^-_{trap}) before the energetically stabilized electron (e^-_{hyd}) is of considerable interest for the understanding of the redox reactions involving univalent reduction. During the coupling of an epithermal electron with water, the very fast appearance of a trapped electron implies that efficient mechanisms do not require large molecular reorganization. The dynamics of electron localization is influenced by the presence of a charged interface (aqueous and reversed micelles). The configurational stability of e^-_{trap} would be achieved by medium energy rearrangement. In the hydration cavity, fast reactions can occur with geminate species such as hydrated proton (H_3O^+) or hydroxyl radical (OH). Considering the isotopic effect on the dynamics of geminate pair recombination, it can be suggested that the energy of vibrational mode of the antisymetric strech can be determinent on the electron-water molecules coupling and the reactivity of transient states of electron (non fully hydrated electron). Femtosecond initiation of a monoelectronic transfer in organized assemblies containing a pyridine nucleotide directly demonstrates that an ultrafast free radical reaction leading to the formation of NAD° occurs with an upper limiting rate of $5\ 10^{11}\ s^{-1}$. This fast redox process being competitive of the electron hydration step would suggest that in biological media, non equilibrium configurations of electron play a dominant role in one electron transfer processes. Recent investigations on reactivity of non fully hydrated electron with disulfide bridges (Gauduel et al., in preparation) permit to develop new concepts on the redox reactions in biology (non equilibrium reactional processes) and to provide experimental basis for testing theoretical works which explore the obvious quantum mechanical character of the structural and dynamical properties of the electron in aqueous media with low or high ionic strength.

<u>REFERENCES</u>

Gauduel, Y., Berrod, S., Migus, A., Yamada, N. and Antonetti, A.(1988[a]) Biochemistry <u>27</u>, 2509-2518.
Gauduel, Y., Migus, A. and Antonetti, A. (1988[b]) In: Chemical reactivity in liquids (M. Moreau, P., Turq, Eds.) Plenum Press, pp 15-32 and references therein.
Gauduel, Y., Pommeret, S., Yamada, N., Migus, A.and Antonetti, A. (1989) J. Am. Chem. Soc. <u>111</u>, 4974-4980.
Jonah, C.D., Miller, J.R. and Matheson, M.S. (1977) J. Phys. Chem. <u>81</u>, 1618-1621.
Maroncelli, M., Castner, E.W., Bagchi, B. and Fleming G.R. (1988) Faraday Discuss. Chem. Soc. <u>85</u>, 199-210.
Migus, A., Antonetti, A., Etchepare, J., Hulin, D. and Orszag, A. (1985) J. Opt. Soc. Am. (B) <u>2</u>, 584-594.
Razem, D., Hamill, W.H. (1977) J. Phys. Chem. <u>81</u>, 1625-1631.
Williams, R.J.P. (1989) Biochem. Int. <u>18</u>, 475-499.

BIOELECTROCHEMICAL PROPERTIES OF GLUTATHIONE.

A NOISE SPECTROGRAPHY INVESTIGATION.

H. Kranck, M-A. Rix-Montel and D. Vasilescu

Biophysics Laboratory, University of Nice - Sophia Antipolis, Parc Valrose
06034 Nice Cedex, France

SUMMARY : Using the noise spectrography technique we have evaluated, in absence of any applied electrical field, the intrinsic electrochemical properties of the reduced monoanionic glutathione at pH 7. GSH^- mechanical mobility determination has led us to represent the reduced monoanionic glutathione as a revolution ellipsoïd of 16 Å and 10 Å axis lengths.

INTRODUCTION

In the cells, the most abundant non-protein thiol is the tripeptide :

γ-L-glutamyl-L-cysteinylglycine or glutathione (GSH). This molecule presents functionnal activities in biology, physiology, cancer therapy and pharmacology (Kosower and Kosower, 1978 ; Meister, 1983 ; Jocelyn, 1972 ; Larsson et al., 1983 ; Mitchell, 1988). Glutathione is generally present inside cells as free sulfhydryl GSH (reduced form) in the concentration range (0.5 to 10 x 10^{-3} M). The disulfide GSSG (oxydized form) is also present in the concentration range (5 to 50 x 10^{-6} M).

At neutral pH, the ionized form of reduced glutathione is :

280

Then, globally, GSH is monoanionic (GSH⁻) at pH 7. On the other hand, this molecule plays an important role in radioprotection, when competing with both DNA and cationic exogenous aminothiols (Fahey et al., 1983). Recently, GSH was also involved in cadmium toxicity (Singhal et al., 1987). So, it appears that the determination of GSH⁻ electrochemical properties is very important with regard to its role at the molecular level. The aim of this work is to determine the electrical and the mechanical mobilities of GSH⁻, in absence of applied electrical field, by using the noise spectrography technique.

MATERIALS AND METHODS

CONTROLS

Free acid glutathione from SIGMA was dissolved in NaCl 10^{-2} M.

Spectrophotometric measurements were made using a Gilford 2400-2 spectrophotometer connected to a 2527 thermoprogrammer. We have studied the variations of optical density with wavelength. A peak was obtained at 195 nm. The Beer-Lambert absorption coefficient ε is 1507 m^2 mol^{-1}.

The pH of the solution was adjusted at 7 by adjunction of NaOH :

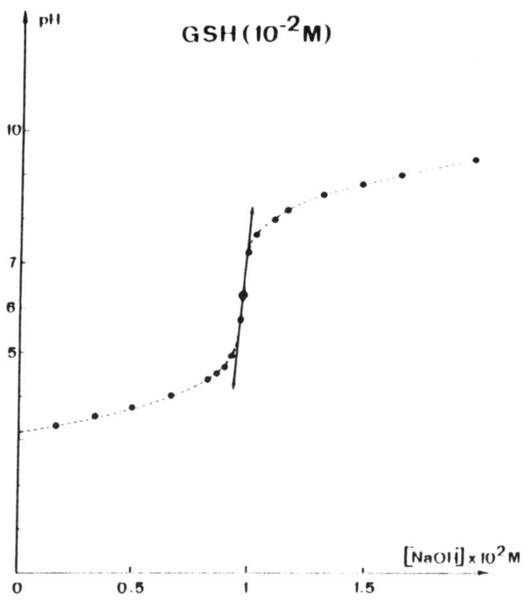

Fig. 1 : Titration curve of GSH (10^{-2} M) in NaCl 10^{-2} M.

At this neutral pH, GSH is globally monoanionic. The titration curve, obtained in our laboratory (Fig. 1), shows that, at pH 7, we add 10^{-2} M NaOH for 10^{-2} M GSH anions. In this case, the electroneutrality is ensured.

METHODS

The ionic conductivity of GSH^- may be estimated by noise measurements at low frequency 5 KHz or by dielectric measurements at high frequency 2 MHz.

Noise Measurements

Brownian motion of charge carriers produces an electrical noise. The sample noise was analysed with a noise spectrograph built in our laboratory. The set up is shown schematically on Fig. 2.

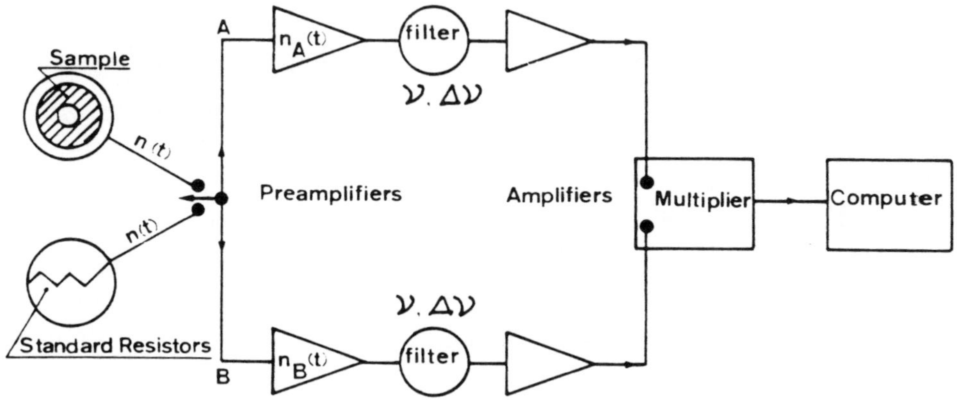

Fig. 2 : Block diagram of the cross-correlation noise spectrograph.

The random voltage n(t) produced by the sample cell or by standard resistors is treated immediately by two identical parallel measurements devices (Vasilescu and Kranck, 1986).

In both channels A and B, n(t) is amplified (frequency range : 0-300 kHz) and filtered at a frequency ν (1 Hz $<\nu<$ 100 kHz) with a bandwidth : $\Delta\nu = 0.05 \, \nu$. If $n_A(t)$ and $n_B(t)$, the two amplifiers proper noises are taken into account, the voltages at the multiplier inputs are :

$$N_A(t) = G\,[n(t) + n_A(t)] \quad \text{and} \quad N_B(t) = G\,[n(t) + n_B(t)]$$

282

where G is the total amplification gain in each channel. The $N_A(t)$ and $N_B(t)$ cross-correlation function $\mathcal{R}_{N_A N_B}(\tau)$ is :

$$\mathcal{R}_{N_A N_B}(\tau) = <N_A(t+\tau) N_B(t)>$$
$$= G^2\left\{ \mathcal{R}_n(\tau) + \mathcal{R}_{n_A n}(\tau) + \mathcal{R}_{nn_B}(\tau) + \mathcal{R}_{n_A n_B}(\tau) \right\}$$

where τ is the correlation time and $\mathcal{R}_n(\tau)$ the autocorrelation function.

The last three cross-correlation functions are negligible because $n_A(t)$ and $n_B(t)$ are non-correlated and independent of $n(t)$. Consequently :

$$\mathcal{R}_{N_A N_B}(\tau) = G^2 \mathcal{R}_n(\tau)$$

By this technique it is possible to eliminate the device's parasitic proper noises. The multiplier-integrator computes the value of $\mathcal{R}_{N_A N_B}(\tau)$ for $\tau = 0$ and the autocorrelation function of the noise produced by the studied sample is :

$$\mathcal{R}_n(0) = G^2 <n(t)^2>$$

In our spectrograph, the sample noise mean-square value $<n(t)^2>$ is compared with the noise produced by standard metal resistors using an interpolation method.

The noise voltage created by a resistor R_n is given by the Nyquist-Johnson theorem :

$$<n^2(t)> = \mathcal{R}_n(0) = 4\,kTR_n\,\Delta\nu$$

where k is the Boltzmann constant, T is the absolute temperature and $\Delta\nu$ the frequency bandwidth.

Measurements were performed at $\nu = 5$ KHz with $\Delta\nu = 250$ Hz. Therefore the resulting noise conductivity is :

$$\sigma_n = g\,R_n^{-1}$$

where g is a geometrical constant.

<u>Impedance Measurements</u>

Complementary dielectric measurements were performed using a Hewlett Packard impedancemeter (HP 4192 A), which allows us to measure the parallel capacitance C_p and the parallel conductance G_p of the samples. This instrument is equiped with a fast automatic frequency scan and driven by an HP 9826 computer. The solutions were enclosed in a cylindrical condenser of geometrical constant g', thermostated at 25°C. We obtained the dielectric conductivity :

$$\sigma_n = \frac{\varepsilon_0 \, G_p}{g'}$$

where ε_0 is the permittivity of the free space.

RESULTS AND DISCUSSION

<u>ELECTRICAL AND MECHANICAL MOBILITIES OF GSH$^-$</u>

The measured noise conductivity σ_n^{mes} of our solutions is expressed by :

$$\sigma_n^{mes} = \sigma_{NaCl} + \sigma_{Na^+} + \sigma_{GSH^-}$$

where σ_{Na^+} is due to NaOH added during adjustement to pH 7.
This equation is justified by the following assertions :

   - GSH$^-$ is the most abundant form at pH 7 ; the H$^+$ and OH$^-$ contributions are negligible at utilised GSH$^-$ concentrations.

   - the electroneutrality is ensured because at pH 7 we have 10^{-2} M NaOH for 10^{-2} M GSH anions.

So, we obtain the noise electrical mobility of GSH$^-$:

$$\mu_{GSH^-} = \frac{\sigma_n}{N \, e}$$

where N is the number of GSH$^-$ ions per m^3 and e the protonic charge (Vasilescu et al., 1974 ; Rix-Montel et al., 1986). From the experimental point of view, Fig. 3 shows that, in our concentration range (0 to 2.4×10^{-2} M) the noise conductivity of GSH$^-$ is linear versus concentration. On the same figure we have reported the results obtained by impedance measurements. They are in harmony with the noise investigation ones.

284

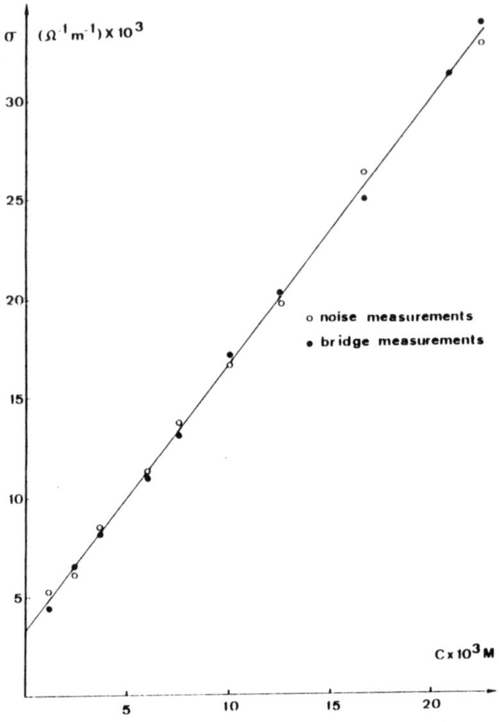

Fig. 3 : Conductivity of GSH⁻ in NaCl 10^{-2} M versus concentration.

So, the experimental value of GSH⁻ electrical mobility is well determined :

$$\mu_{GSH^-} = (1.57 \pm 0.08) \times 10^{-8}\, m^2\, s^{-1}\, V^{-1}$$

We can establish that, for a concentration 10^{-2} M and at 25°C :

$$\mu_{GSH^-} \simeq \frac{\mu_{Na^+}}{3}$$

PERRIN'S ELLIPSOID MODELISATION OF GSH⁻

From the electrical mobility we can obtain the mechanical mobility of the monoanionic glutathione :

$$u_{GSH^-} = \frac{\mu_{GSH^-}}{e}$$

Our experimental value is $u = 9.8 \times 10^{10}\, kg^{-1}\, s$.

If GSH⁻ is considered as an ellipsoid whose semi-axis lengths are a and b (a > b), the diffusion coefficient D associated with translational brownian motion is, using Perrin's formulation (Perrin, 1934 ; Perrin, 1936) :

$$D = \frac{kTS}{12\pi\eta}$$

where D is expressed in $m^2 s^{-1}$, S in m^{-1} and η (water viscosity coefficient) is 0.89×10^{-3} Kg m^{-1} s^{-1} at 25°C with :

$$S = \frac{2}{a(1-p^2)^{1/2}} \ln \left\{ \frac{1 + (1-p^2)^{1/2}}{p} \right\} \quad \text{where } p = b/a$$

Then using the Einstein relation D = kTu, we obtain the theoretical values of the ellipsoid mechanical mobility :

$$u_{th} = \frac{S}{12\pi\eta}$$

Taking into account GSH$^-$ CPK molecular models we have computed theoretical values of mechanical mobilities for different semi-axis lengths :

$$7 \overset{o}{A} < a < 9 \overset{o}{A} \quad \text{and} \quad 4 \overset{o}{A} < b < 6 \overset{o}{A}$$

The best agreement between theoretical and experimental values is obtained for an ellipsoid of semi-axis lengths : a = 8.5 $\overset{o}{A}$ and b = 5 $\overset{o}{A}$ which gives :

$$u_{th} = 9.75 \times 10^{10} \text{ kg}^{-1} \text{ s}$$

It is to be noted that b variations have small effects on u_{th} values. Nevertheless an ellipsoid of a = 8 $\overset{o}{A}$ is in best agreement with experimental and theoretical conformationnal results (Laurence and Thomson, 1980 ; York et al., 1987 ; Podanyi and Reid, 1988). In this case we find :
u_{th} = 9.99 $\times 10^{10}$ kg^{-1} s which is not very far from experimental u value.

In conclusion the noise determination of GSH$^-$ mechanical mobility has led us to represent the reduced monoanionic glutathione as a revolution ellipsoid of 16 $\overset{o}{A}$ and 10 $\overset{o}{A}$ axis lengths. This ellipsoid contains the "pseudo random" conformation of GSH$^-$ molecule at pH 7 (see figure 4).

286

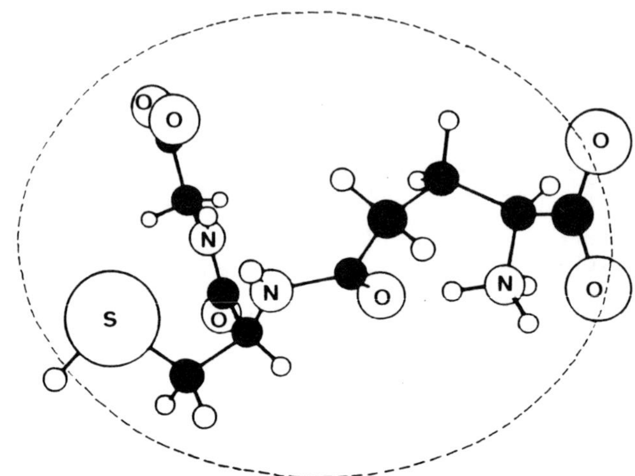

Fig. 4 : Ellipsoidal modelisation of GSH⁻ molecule. The represented
form is the optimum PCILO computed conformation obtained
by P.R. Laurence & C. Thomson.

Aknowledgments : We wish to thank Dr. H. Sentenac-Roumanou and Dr. M.
Nouaille-Degorce for fruitful suggestions. This research was supported
by grants DRET (projects 86/114 and 89/089).

REFERENCES

Fahey, R.C., Dorian, R., Newton, G.L. and Utley, J. (1983) in : Radioprotec-
    tors and anticarcinogens, (Nygaard, O.F. and Simic, M.G., Eds) Academic
    Press, New York, pp. 103.
Jocelyn, P.C. (1972) "Biochemistry of the SH groups", Acad. Press, N.Y.
Kosower, N.S. and Kosower, E.M. (1978) Int. Rev. Cytol. 54, 109-160.
Larsson, A., Orrenius, S., Holmgren, A. and Mannervik, B. (1983) Functions
    of Glutathione, Raven Press, N.Y.
Laurence, P.R. and Thomson, C. (1980) Theoret. Chem. Acta 57, 25-41.
Meister, A. (1983) Science 220, 472-477.
Mitchell, J.B. (1988) Isi Atlas of Science : Pharmacology, 155-160.
Perrin, F. (1934) J. Phys. Radium 5, 497-511.
Perrin, F. (1936) J. Phys. Radium 7, 1-11.
Podanyi, B. and Reid, R.S. (1988) J. Am. Chem. Soc. 110, 3805-3810.
Rix-Montel, M.A., Kranck, H. and Vasilescu, D. (1986) Bioelectrochemistry
    and Bioenergetics 16, 427-434.
Singhal, R.K., Anderson, M.E. and Meister, A. (1987) FASEB J. 1, 220-223.
Vasilescu, D. and Kranck, H. (1986) in : Modern Bioelectrochemistry (F.
    Gutmann and H. Keyser, Eds), Plenum Press, N.Y., p. 397.
Vasilescu, D., Teboul, M., Kranck, H. and Gutmann, F. (1974) Electrochem.
    Acta 19, 181-186.
York, M.J., Beilharz, G.R. and Kuchel, P.W. (1987) Int. J. Peptide Protein
    Res. 29, 638-646.

TAKING INTO ACCOUNT THE CELL WATER PROPERTIES
FOR THE CYTOCHEMICAL DETECTION OF CATIONS:
EMBEDDING INTO MELAMIN AFTER PYROANTIMONATE FIXATION.

P. Mentré

C.N.R.S., 67 rue Maurice-Gunsbourg, 94200 Ivry-sur-Seine, France.

Material, fixed in presence of K pyroantimonate, was dehydrated in
a very graded series of alcohols before embedding in an hydroxy
resin, or was directly embedded in a hydrosoluble resin. The cation
visualization was improved, particularly in the second case.

The pyroantimonate (PA) method consists in the fixation of
biologic material in presence of potassium pyroantimonate in order
to visualize the cation distribution with electron microscope, as
electron opaque precipitates (Komnick, 1962). The Na^+, Mg^{++} and Ca^{++}
free cations precipitate at the concentrations of 10^{-2}, 10^{-5} and
10^{-6} M respectively (Klein et al., 1972). We have previously
proposed an adaptation of this method, completed by electron probe
microanalysis, demonstrating that the complexed cations, according
to the accessibility of their charges can be PA 'stained' or remain
'masked' (Mentré & Halpern, 1988). Contrary to the widely held
opinion, we did not observed that great distance ion diffusion
occurs during the procedure. Nevertheless, some structures which we
expected to be finely stained, for example the intercellular spaces,
appeared visualized with coarse precipitates.

For the electron microscopy, the material is routinely embedded
in an epoxy resin, which requires a previous alcoholic dehydration.

We thought that this dehydration which breaks the H-bonds binding water to macromolecules, could produce 'microaffluxes' of pure water at the molecular scale and dissolve the smallest PA-cation complexes. Therefore, we have assayed to reduce these affluxes of water either in proceeding to a more progressive alcoholic dehydration or in embedding our material directly in a hydrosoluble resin.

MATERIAL AND METHODS

Different organs of mouse, rat and quail were fixed for 2 hours at 4°C in presence of 4% potassium pyroantimonate and 2% paraformaldehyde, as previously described (Mentré & Halpern, 1988).

For embedding in the epoxy resin Araldite, the fixed material was rinsed or not with distilled water and dehydrated in a series of alcohols: 25°-D (25-50-70-95-100°); or 50°-D (50-70-95-100°); or 70°-D (70-95-100°). Each alcohol bath was of 10 min. Ethanol was replaced with propylene oxide and embedding in Araldite.

For embedding in the hydrosoluble resin Melamin, the fixed material was briefly wiped and directly embedded in a medium containing 10 g of Melamin MME 7002 and 0.02 g of catalyst acid (Nanoplast FB101, Agar Aids). According to the instructions of the manufacturer, the material was dried for 2 days at 40°C and polymerized for 2 days at 60°C.

Ultrathin sections of the embedded material were observed with an electron microscope Philips EMU 300 at 80 kV.

RESULTS

The results were very similar in all the studied tissues.

Alcoholic dehydration and embedding in Araldite: A careful rinse after the fixation practically extracted the whole pyroantimonate. A brief rinse of a few seconds (comparison with not rinsed controls) permitted to wash off the fixative retained outside of the cells without modifying the PA distribution inside of the cells. After

70°-D, some structures were finely stained, for example the glycogen granules in glycogenolysis (Mentré & Halpern, 1989). Some cells retained PA more intensively than others. But the intercellular spaces were coarsely marked with precipitates and the plasma membranes were never stained. After 50°-D, the intercellular spaces and the basal laminae were generally clearly PA delineated. The plasma membranes were rarely stained. Glycogen and endocytotic

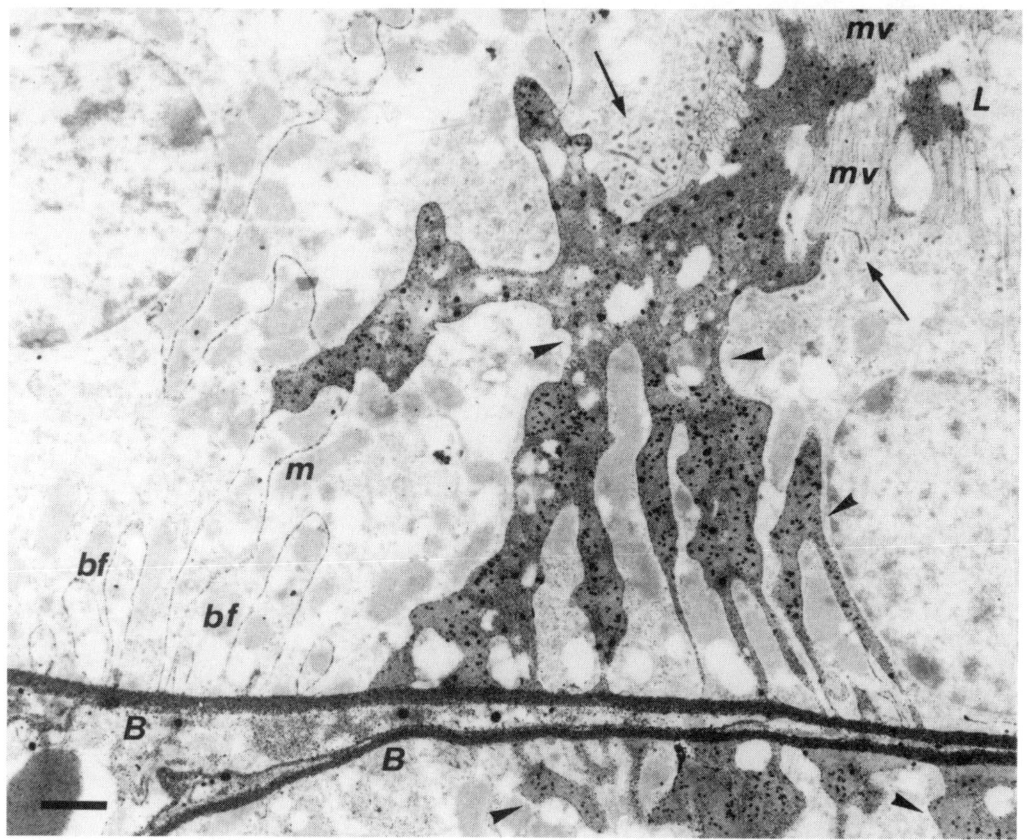

Figure 1. Proximal convoluted tubule in mouse kidney after 50°-D dehydration and embedding in Araldite. The cells are clearly delineated by PA.and the basal laminae (B) very contrasted. Numerous basal folds (bf), enclosing mitochondria (m) are visible. On the apical side, facing the lumen (L) of the tubule, the microvilli (mv) are recognizable, but they are generally not finely stained. Cation-rich endocytotic canaliculi (arrows), well contrasted, penetrates into the cells. Some cells (arrowheads), morphologically identical to the others, contain numerous PA precipitates. Bar = 1 μm.

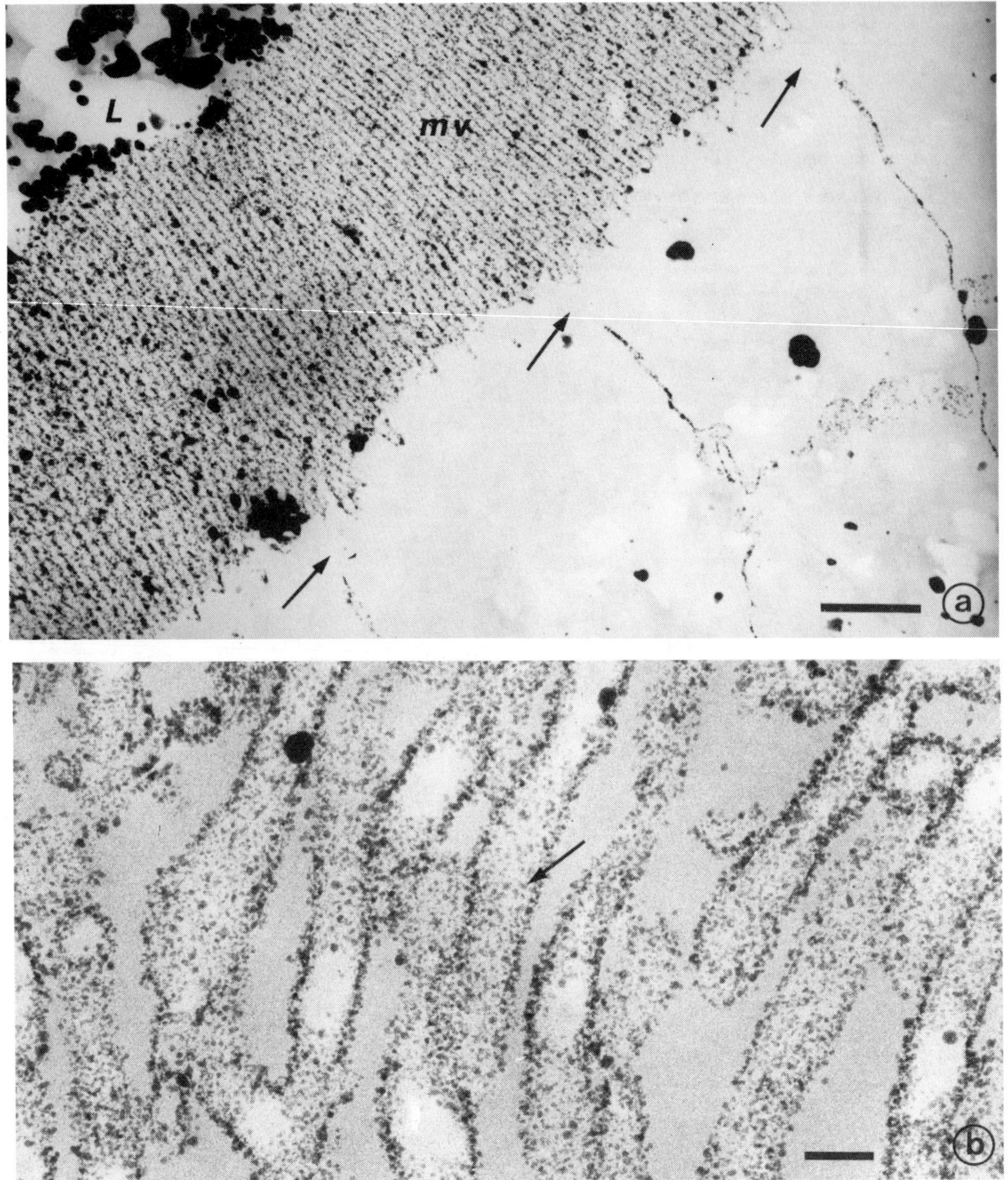

Figure 2. Enterocytes in quail small intestine, after embedding in Melamin. a: The enterocytes are clearly PA delineated excepted at the level of the tight junctions (arrows). The microvilli (mv), along the lumen (L) are densely loaded with PA precipitates. Bar=1μm. b: Detail of microvilli. Numerous particles, finely stained, about 10nm in diameter, are regularly arranged on the microvilli surface (arrow). Bar=0.1μm.

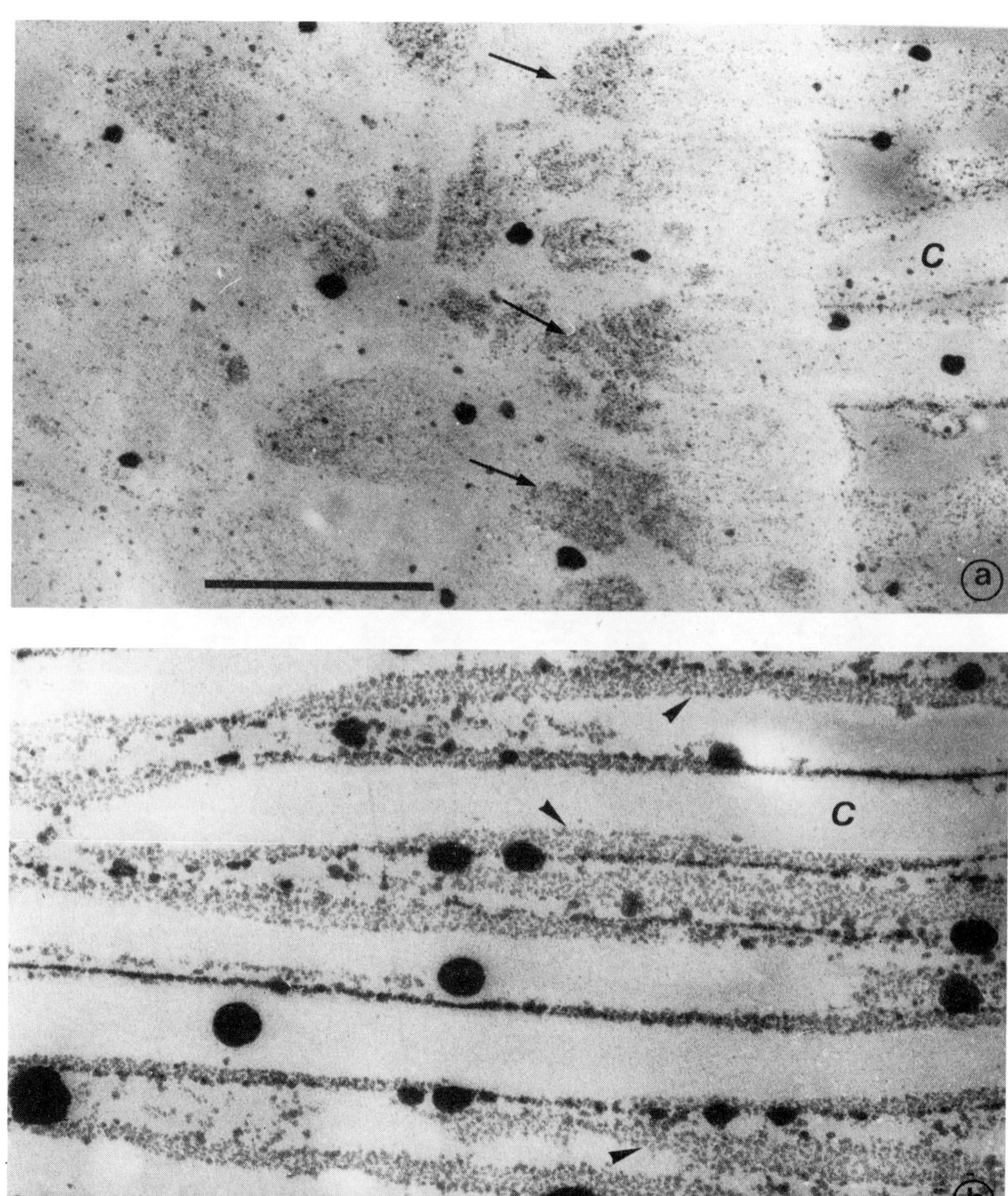

Figure 3. Ciliated epithelium of quail trachea embedded in Melamin. a: the ciliary roots (arrows) located at the base of the cilia (C) are finely stained by PA. b: the surface of the cilia is covered with particles, about 10nm in diameter, finely stained and regularly arranged. Bar = 0.5 μm.

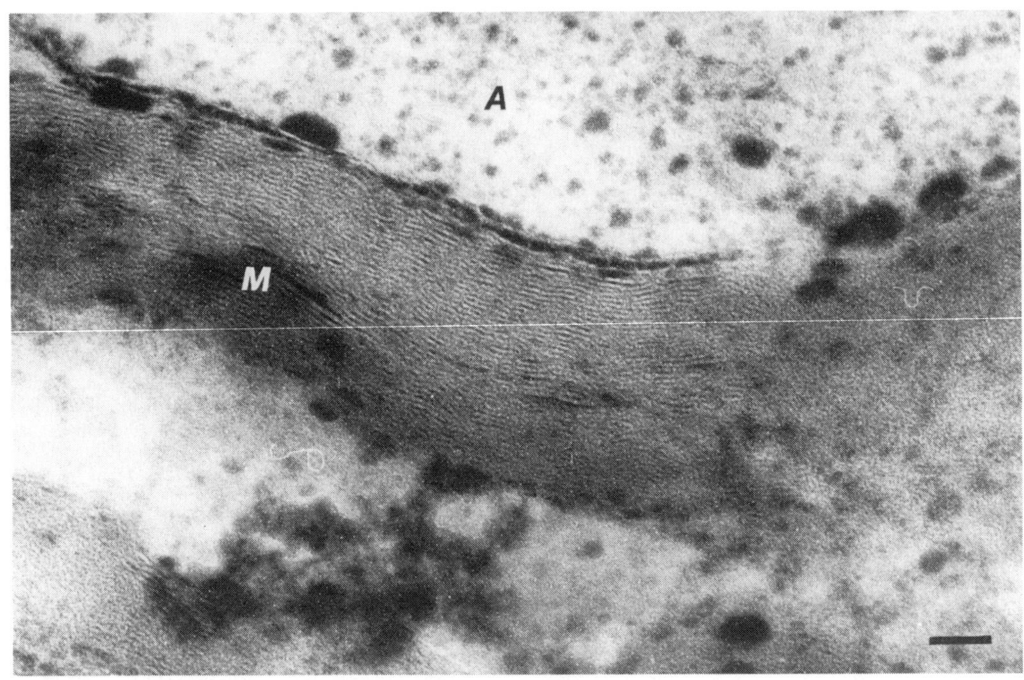

Figure 4. White substance in rat cerebellum embedded in Melamin. A myelinated axon (A) with layers of myelin (M) clearly delineated by pyroantimonate. Bar = 0.1μm.

vesicles appeared well contrasted. Some cells were more contrasted than others, as the secretory Paneth cells in the small intestine and a few cells in the proximal convoluted tubule of kidney. Representative results are shown (Fig. 1.). After 25°-D, as after rinse, nearly all the PA.was extracted.

Embedding into Melamin: The intercellular spaces were clearly PA delineated (Fig. 2a). Various intracellular structures were stained, as glycogen, secretory granules, ciliary roots (Fig. 3a)..Finely stained particles, about 10 nm, uniformly arranged, were observed on plasma membranes, for example the microvilli in small intestine (Fig. 2b) and the cilia in trachea (Fig. 3b). The Figure 4 illustrates the fine contrast observed in the layers of myelin of the neuron myelin sheaths. In most of the cases, this staining was very instable, disappearing if the material was rinsed, even shortly, after the fixation. But without rinse, it was remarkably reproducible and precise. Nevertheless, it is worth noting that the penetration of Melanin into the material - and consequently the embedding and the quality of the sections - was not excellent beyond 100 μm. Moreover, coarse PA precipitates, apparently distributed at random, were abundant (Fig. 2, 3 and 4).

DISCUSSION

The PA method provides information about the distribution and the state (free, easily liberated or bound) of cations (Mentré & Halpern, 1988). Our observations appear to be in agreement with already known results: the richness in bound calcium of the extracellular matrices - intercellular cement and basal laminae - (Weiss, 1984), of the ciliary roots (Salisbury, 1984); the affinity for calcium and magnesium of the myelin due to its richness in phospholipids (Lederer, 1987); the richness in cation channels, transporters and pumps of the apical plasma membranes (Alberts at al., 1983); the increase of free cations in the metabolic active cells (Lederer, 1987). The presence, in the proximal tubule of kidney, of cells morphologically identical to the others but chemically different (Fig. 1), could be related to the fact that cytochemists have identified several kinds of cells differing by enzymatic activities (Longley, 1969).

Our results may be explained by the three-dimensional configuration of water in the cell proposed in 1988 by Cameron et al.; Hazlewood & Kellermayer; Ling; Negendank. A very low fraction of water would be free, therefore available for the diffusion of the ions and small uncharged molecules. Moreover, the ions would be trapped by many macromolecules. In these conditions, even during the fixation (Mentré & Halpern, 1989), the ions, would not displace upon long distances. During 70°-D, bound water would be massively liberated from macromolecules, dissolving the smallest PA-cation complexes. This dissolution would be immediately followed by precipitation because the pyroantimonates are very insoluble in alcohol. With 25°-D, the PA complexes are washed off, as with rinse.

The Melamin initially contains 30% water. During the dessication at 40°C, the evaporation gradient carries away the water of the material, permitting its exchange for Melamin. The excess of PA would be the cause of the randomly distributed precipitates. But this excess of reagent may be necessary (as in many chemical reactions) to preserve the reproducible tiny complexes which are lost otherwise.

In practice, embedding in Melamin should be useful in the cases where a precise distribution is required. But it does not permit to obtain easily routine ultrathin sections. Consequently, in the other cases, Araldite, following 50°-D, would be more convenient.

Aknowledgements: The discussion which followed the symposium on Potassium at the 'Scanning Microscope/89' meeting have been for us a source of ideas. We wish to express our gratitude to the Drs. R.T. Kado ,C. Batini and G. Nicaise for their kind comments and suggestions and to M. Louette for his excellent glossy prints.

REFERENCES

Alberts B., Bray D., Lewis J., Raff M., Roberts K., and Watson J.D. (1983) In: Molecular Biology of the Cell, Garland Publishing, Inc., New-York, pp. 286-302.
Cameron I.L., Fullerton G.D., and Smith N.K.R. (1988) Scanning Microsc. 2, 275-288.
Hazlewood C.F., and Kellermayer M. (1988) Scanning Microsc. 2, 267-273.
Klein R.L., Yen S.S., and Thureson-Klein A. (1972) J. Histochem. Cytochem. 20, 65-78.
Komnick H. (1962) Protoplasma 55, 414-418.
Lederer J. (1984) In Magnésium, Mythes et Réalités (Nauwelaerts , Eds), Maloine, Paris, pp. 26-36.
Ling G.N. (1988) Scanning Microsc. 2, 871-884.
Longley J.B. (1969) In The Kidney (C. Rouiller and A.F. Muller, eds), Academic Press, New-York, 157-259.
Mentré P., and Halpern S. (1988) J. Histochem. Cytochem. 36, 55-64.
Mentré P., and Halpern S. (1989) Scanning Microsc. 3, in press.
Negendank W. (1988) Scanning Microsc. 2, 21-32.
Salisbury J.L., Baron A., Surek B., and Melkonian M. (1984) J. Cell Biol. 99, 962-970.
Weiss L. (1983) In: Matrices and Cell Differentiation (R.B. Kemp and J.R. Hinchliffe, Eds), A.R. Liss, New-York, pp. 371-386.

SUBJECT INDEX
Page numbers refer to first page of each contribution